4

GUIDANCE NOTE 4

Protection Against Fire

18th IET Wiring Regulations BS 7671:2018

Published by the Institution of Engineering and Technology, London, United Kingdom

The Institution of Engineering and Technology is registered as a Charity in England & Wales (no. 211014) and Scotland (no. SC038698).

 The Institution of Engineering and Technology is the institution formed by the joining together of the IEE (The Institution of Electrical Engineers) and the IIE (The Institution of Incorporated Engineers).

First published 1992 (0 85296 539 7)

Reprinted (with amendments) 1993

Reprinted (with minor amendments) 1994

Second edition (incorporating Amendment No. 1 to BS 7671:1992) 1995 (0 85296 868 X)

Third edition (incorporating Amendment No. 2 to BS 7671:1992) 1998 (0 85296 957 0)

Fourth edition (incorporating Amendment No. 1 to BS 7671:2001) 2003 (0 85296 992 9)

Reprinted (incorporating Amendment No. 2 to BS 7671:2001) 2004

Fifth edition (incorporating BS 7671:2008) 2009 (978-0-86341-858-7)

Reprinted 2010

Sixth edition (incorporating Amendment No. 1 to BS 7671:2008) 2012 (978-1-84919-277-4)

Reprinted 2012, 2014

Seventh edition (incorporating Amendment Nos. 2 and 3 to BS 7671:2008) 2015 (978-1-84919-875-2)

Eighth edition (incorporating 18th Edition to BS 7671:2018) 2018 (978-1-78561-455-2)

Copies of this publication may be obtained from:

The Institution of Engineering and Technology
PO Box 96, Stevenage, SG1 2SD, UK
Tel: +44 (0)1438 767328
Email: sales@theiet.org
www.electrical.theiet.org/books

ISBN 978-1-78561-455-2 (paperback)
ISBN 978-1-78561-456-9 (electronic)

Typeset in the UK by the Institution of Engineering and Technology, Stevenage
Printed in the UK by Sterling Press Ltd, Kettering

Contents

Cooperating organisations 6

Acknowledgements 7

Preface 8

Introduction 9

Chapter 1   Statutory requirements 11

1.1   General 11
1.2   Statutory Regulations 11
    1.2.1   The Electricity at Work Regulations 1989 as amended 11
    1.2.2   The Electricity Safety, Quality and Continuity Regulations 2002 as amended 11
    1.2.3   The Management of Health and Safety at Work Regulations 1999 as amended 12
    1.2.4   The Provision and Use of Work Equipment Regulations 1998 12
    1.2.5   The Construction (Design and Management) Regulations 2015 12
    1.2.6   The Dangerous Substances and Explosive Atmospheres Regulations 2002 as amended 12
    1.2.7   The Petroleum (Consolidation) Regulations 2014 13
    1.2.8   The Cinematograph (Safety) Regulations 1955 as amended 13
    1.2.9   The Building Regulations 2010 (England and Wales) as amended 13
    1.2.10   The Building (Scotland) Regulations 2004 as amended 13
    1.2.11   The Building Regulations (Northern Ireland) 2012 as amended 13
    1.2.12   The Health and Safety (Safety Signs and Signals) Regulations 1996 13
    1.2.13   The Regulatory Reform (Fire Safety) Order 2005 (England and Wales) 14
    1.2.14   Other Regulations 15

Chapter 2   Overview of the Wiring Regulations 17

Chapter 3   Generally applicable requirements 21

3.1   General 21
3.2   Fixed equipment 22
    3.2.1   Heating cables 22
    3.2.2   Luminaires 22
    3.2.3   Extra-low voltage lighting installations 22
    3.2.4   Symbols used in connection with luminaires, controlgear for luminaires and their installation 22
3.3   Manufacturers' instructions 24

3.4 No fire risk in normal use 24
3.5 Arcs, sparks and hot particles 27
3.6 Protection against arcing 27
3.7 Focusing and concentration of heat 28
3.8 Equipment containing flammable liquid 28
3.9 Enclosures 30
    3.9.1 Consumer units and similar assemblies in
        domestic premises 30
3.10 Terminations 31
3.11 Installation of Cables 31

Chapter 4   Precautions where particular risks of fire exist   33

4.1 General 33
4.2 Requirements for emergency escape routes 33
4.3 Risks of fire due to the nature of processed or stored materials 34
    4.3.1 Luminaires 35
    4.3.2 Enclosures 35
    4.3.3 Switchgear and controlgear 35
    4.3.4 Cables, circuits and wiring systems 36
    4.3.5 Motors 37
    4.3.6 Heating appliances 38
4.4 Combustible constructional materials 38
    4.4.1 Equipment 38
    4.4.2 Luminaires 38
    4.4.3 Cables and wiring systems 39
4.5 Fire propagating structures 39
4.6 Locations of national, commercial, industrial or public significance 39

Chapter 5   Protection against burns   41

Chapter 6   Measures to minimise the spread of fire   43

6.1 Precautions in a fire-segregated compartment 43
6.2 Sealing of wiring system penetrations 44
    6.2.1 External sealing arrangements 44
    6.2.2 Internal sealing arrangements 45
    6.2.3 Material for sealing wiring system penetrations 45
    6.2.4 Flush-mounted accessories penetrating stud partitions 46
    6.2.5 Inspection 48

Chapter 7   Safety services   49

7.1 General 49
7.2 Classification and changeover times 49
7.3 Sources for safety services 50
    7.3.1 Generators 50
7.4 Circuits 52
7.5 Information to be made available 53
7.6 Wiring systems 53

Chapter 8   Special installations and locations   57

8.1 Introduction 57
8.2 Agricultural and horticultural premises 57
    8.2.1 The risks 57
    8.2.2 Radiant heaters 58
    8.2.3 Employment of RCDs as a measure for protection against fire 58
    8.2.4 Additional requirements for ELV circuits 58

	8.2.5	Rodent activity	58
	8.2.6	Luminaires	59
8.3		Medical locations	59
8.4		Solar photovoltaic (PV) power supply systems	59
8.5		Extra-low voltage lighting installations	60
8.6		Exhibitions, shows and stands	61
8.7		Temporary electrical installations for structures, amusement devices and booths at fairgrounds, amusement parks and circuses	61
8.8		Heating cables and embedded heating systems	61

Appendix A	**Safety service and product standards of relevance to this Guidance Note**	**63**
Appendix B	**Approved Document B1: Means of warning and escape**	**69**
Appendix C	**Approved Document B2: Internal fire spread (linings)**	**83**
Appendix D	**Approved Document B3: Internal fire spread (structure)**	
Appendix E	**Scottish Technical Standards relating to protection against fire**	**87**
Index		**105**

Cooperating organisations

The Institution of Engineering and Technology acknowledges the invaluable contribution made by the following individuals in the preparation of this Guidance Note.

Institution of Engineering and Technology

M. Coles BEng(Hons) MIET
P.E. Donnachie BSc CEng FIET
Eur Ing L. Markwell MSc BSc(Hons) CEng MIEE MCIBSE LCGI
G.G. Willard DipEE CEng MIEE JP

We would like to thank the following organisations for their continued support:

BEAMA Ltd
Certsure trading as NICEIC and Elecsa
ECA
Electrical Contractors' Association of Scotland (SELECT)
Health and Safety Executive
NAPIT

Guidance Note 4 revised, compiled and edited

T. R. Pickard IEng MIET and Ewhen Kochan Msc, BSC(Hons), MIET, AMCIBSE, MSET.

Acknowledgements

References to British Standards, CENELEC Harmonization Documents and International Electrotechnical Commission standards are made with the kind permission of BSI. Complete copies can be obtained by post from:

BSI Customer Services
389 Chiswick High Road
London W4 4AL
Tel: +44 (0)20 8996 9001
Fax: +44 (0)20 8996 7001
Email: cservices@bsigroup.com

BSI also maintains stocks of international and foreign standards, with many English translations. Up-to-date information on BSI standards can be obtained from the BSI website: www.bsi-group.com

Complete copies of Approved Document B and the other Approved Documents are downloadable from the planning portal website: www.planningportal.gov.uk

The Building Standards Division of the Scottish Government, Directorate for the Built Environment, website can be found at: https://beta.gov.scot

The Northern Ireland Building Regulations website can be found at: www.buildingcontrol-ni.com/regulations

The illustrations within this publication were provided by Rod Farquhar Design: www.rodfarquhar.co.uk

Cover design and illustration were created by The Page Design: www.thepagedesign.co.uk

The flameguard downlighter and insulation support box images were supplied courtesy of CLICK Scolmore: www.scolmore.com

Copies of Health and Safety Executive documents and approved codes of practice (ACOP) can be obtained from:

HSE Books PO Box
1999 Sudbury, Suffolk
CO10 2WA
Tel: +44 (0)1787 881165
Email: hsebooks@prolog.uk.com
Web: http://books.hse.gov.uk

Preface

This Guidance Note is part of a series issued by the Institution of Engineering and Technology to explain and enlarge upon the requirements in BS 7671:2018, the 18th Edition of the IET Wiring Regulations.

Note that use of this Guidance Note does not ensure compliance with BS 7671. Its content is intended to explain some of the requirements of BS 7671 but readers should always consult BS 7671 to satisfy themselves of compliance.

The scope generally follows that of BS 7671; the relevant Regulations and Appendices are noted in the margin. Some Guidance Notes also contain material not included in BS 7671:2018 but which was included in earlier editions of the Wiring Regulations. All of the Guidance Notes contain references to other relevant sources of information.

Electrical installations in the United Kingdom that comply with BS 7671 are likely to satisfy the relevant parts of Statutory Regulations such as the Electricity at Work Regulations 1989. However, this cannot be guaranteed. It is stressed that it is essential to establish which Statutory and other Regulations apply and to install accordingly. For example, an installation in premises subject to licensing may have requirements different from or additional to those of BS 7671, and these will take precedence.

Introduction

Protection against fire resulting from the installation and use of electrical installations has been necessary ever since the use of electricity was first introduced into buildings. In fact, it is worth noting that the first edition of the *Wiring Regulations*, introduced in 1882, was called the *Rules and Regulations for the Prevention of Fire Risks Arising from Electric Lighting*.

This Guidance Note is concerned primarily with Chapter 42 and those other parts of BS 7671 which concern protection against thermal effects. It does not attempt to deal comprehensively with the safety of personnel in the event of fire, since this is beyond the scope of BS 7671.

Neither BS 7671 nor the Guidance Notes are design guides. It is essential to prepare a full design and specification prior to commencement or alteration of an electrical installation. Compliance with the relevant standards should be required.

514.9 The design and specification should set out the requirements and provide sufficient information to enable competent persons to carry out the installation and to commission it. The specification must include a description of how the system is to operate and all the design and operational parameters. It must provide for all the commissioning procedures that will be required and for the provision of adequate information to the user. This will be by means of an operational and maintenance manual or schedule, and 'as fitted' drawings if necessary.

It must be noted that it is a matter of contract as to which person or organisation is responsible for the production of the parts of the design, specification, construction and verification of the installation and any operational information.

The persons or organisations who may be concerned in the preparation of the works include:

The Designer
The CDM Principle Designer and Principle Contractor
The Installer
The Electricity Distributor and/or Supplier
The Installation Owner (Client) and/or User
The Architect
The Fire Risk Assessor
Any Regulatory Authorities
Any Licensing Authority
The Health and Safety Executive.

132.1 In producing the design, advice should be sought from the installation owner and/or user as to the intended use. Often, as in a speculative building, the intended use is unknown. The specification and/or the operational manual must set out the basis of use for which the installation is suitable.

133.1
Sect 511 Precise details of each item of equipment should be obtained from the manufacturer and/or supplier and compliance with appropriate standards confirmed.

The operational manual must include a description of how the system as installed is to operate and all commissioning records. The manual should also include manufacturers' technical data for all items of switchgear, luminaires, accessories, etc. and any special instructions that may be needed.

The Health and Safety at Work etc. Act 1974 Section 6 and the Construction (Design and Management) Regulations 2015 are concerned with the provision of information, and guidance on the preparation of technical manuals is given in BS EN 82079-1 *Preparation of instructions for use. Structuring, content and presentation. General principles and detailed requirement*s and BS 4940 series *Technical information on construction products and services*. The size and complexity of the installation will dictate the nature and extent of the manual.

As a reminder of what can happen when things go wrong, the case below highlights a tragedy caused by a faulty electrical installation.

Faulty electrical installations can initiate fires and even a seemingly minor defect in an installation can lead to fire. .

During 2004 a fire in a residential care home was initiated by an overloaded cable that was poorly installed at its entry to a distribution board. This defect remained undetected. An electrical fire started and this triggered a general fire. This in turn, coupled with other unfortunate events (including poor evacuation of patients and poor smoke control), resulted in the deaths of 14 persons resident at the home.

The electrical fire was initiated by an overloaded cable. The cable was incorrectly terminated as it entered a metal distribution board, via an access hole to which a grommet was not fitted. Also, the outer sheath of the cable was stripped back outside the distribution board. The cable became thermally stressed and electric arcing occurred at the cable entry position. The arcing then acted as a source of ignition to the general fire.

If a grommet had been fitted at the access hole and the cable sheath had been terminated correctly within the distribution board, then any arcing may well have been contained within the distribution board enclosure.

Statutory requirements

1.1 General

A number of enactments and statutory instruments, including the Electricity at Work Regulations 1989 (made under the Health and Safety at Work etc. Act 1974), deal with fire risks, prevention of fire and fire precautions.

110.1.3 This Guidance Note is not intended to provide an exhaustive treatment of the legislation concerned with fire, but deals only with those situations referred to in BS 7671. Thus, certain specialised installations listed in Regulation 110.1.3 are excluded.

1.2 Statutory Regulations

1.2.1 The Electricity at Work Regulations 1989 as amended

The Electricity at Work Regulations are general in their application and refer throughout to 'danger' and 'injury'. Danger is defined as risk of 'injury' and injury is defined in terms of certain classes of potential harm to persons. Injury is stated to mean death or injury to persons from:

(a) electric shock;
(b) electric burn;
(c) electrical explosion or arcing; and
(d) fire or explosion initiated by electrical energy.

The Health and Safety Executive book *The Electricity at Work Regulations 1989 Guidance on Regulations* (HSR25) is essential reading for all concerned with electrical installations.

1.2.2 The Electricity Safety, Quality and Continuity Regulations 2002 as amended

The definition of 'danger' in these Regulations includes 'danger to health or danger to life or limb from electric shock, burn, injury or mechanical movement to persons, livestock or domestic animals, or from fire or explosion, attendant upon the generation, transmission, transformation, distribution or use of energy'.

1.2.3 The Management of Health and Safety at Work Regulations 1999 as amended

These Regulations place general duties on employers to assess risks to the health and safety of employees and others and take managerial action to minimise these risks, including:

(a) implementing preventive measures;
(b) providing health surveillance;
(c) appointing competent people;
(d) setting up procedures;
(e) providing information; and
(f) training etc.

Note: These Regulations have been subject to some revocations under the Regulatory Reform (Fire Safety) Order 2005. See Schedule 5 of the Order for full details.

1.2.4 The Provision and Use of Work Equipment Regulations 1998

These Regulations require employers to ensure that work equipment is suitable for the purpose.

The Regulations also require that work equipment is maintained in an efficient state, in efficient working order and in good repair.

1.2.5 The Construction (Design and Management) Regulations 2015

The Construction (Design and Management) Regulations (CDM Regulations) require active planning, coordination and management of the building works, including the electrical installation, to ensure that hazards associated with the construction, maintenance and perhaps demolition of the installation are given due consideration, as well as provision for safety in normal use.

Regulation 12 (5) requires a record known as 'the health and safety file' to be prepared, reviewed and updated during the construction process. This file should be passed to the client on completion of the construction work.

Regulation 12 (6) requires the principal designer to review the health and safety file and update and revise as necessary to take account of the work and any changes that have occurred throughout the project. At the end of the project, Regulation 12 (10) requires the principal designer, or where there is no principal designer the principal contractor, to pass the health and safety file to the client. The role of the CDM co-ordinator that was a requirement of the 2007 regulations was transferred to the principle designer and principle contractor in the 2015 regulations.

See CDM Regulations 2015 Part 2 Reg 5(1) "Appointment of the principal designer and the principal contractor".

1.2.6 The Dangerous Substances and Explosive Atmospheres Regulations 2002 as amended

These Regulations place a duty on the employer to assess the risks arising from dangerous substances in the workplace. These include the presence of flammable atmospheres and potential ignition sources. If such conditions cannot be eliminated, the regulations require the selection of electrical equipment appropriate for the areas in which it is to be used that has protection against becoming a source of ignition.

1.2.7 The Petroleum (Consolidation) Regulations 2014

These regulations which came into force on 1st October 2014 apply to retail and non retail petrol filling stations and non workplace premises storing petrol such as private homes or clubs.

For electrical installations in petrol filling stations, guidance can be found in *Guidance for Design, Construction, Modification, Maintenance and Decommissioning of Filling Stations* (known as *The Blue Book*), published jointly by the Energy Institute and the Association for Petroleum and Explosives Administration (APEA), www.apea.org.uk

Guidance on the safe use of petroleum in garages is available on the Health and Safety Executive's website: http://www.hse.gov.uk/pubns/indg331.pdf.

1.2.8 The Cinematograph (Safety) Regulations 1955 as amended

These Regulations apply to premises used for cinematography exhibitions.

1.2.9 The Building Regulations 2010 (England and Wales) as amended

The Building Regulations 2010 cover the construction and extension of buildings and are supported by Approved Documents which set out detailed practical guidance on compliance. Approved Document B 2006 was issued by the Department for Communities and Local Government (CLG) to provide practical guidance on meeting the requirements of Schedule 1 (means of escape) and Regulation 7 (materials and workmanship) of the Building Regulations.

Approved Document B consists of two parts as follows:

(a) Volume 1 – Dwellinghouses (amended 2010 and 2013)
(b) Volume 2 – Buildings other than dwellinghouses (amended 2010 and 2013)

See Appendices B, C and D of this Guidance Note for further information.

1.2.10 The Building (Scotland) Regulations 2004 as amended

Compliance with the Scottish Building Regulations can be achieved by compliance with the mandatory Building Standards and associated guidance given in the two Scottish Building Standards Technical Handbooks (Domestic and Non-Domestic).

See Appendix E of this Guidance Note for further information.

1.2.11 The Building Regulations (Northern Ireland) 2012 as amended

The Department of Finance and Personnel has produced *Technical Booklet E:2005* and also refers to other deemed-to-satisfy publications, such as British Standards, to support compliancewith Part E (Fire safety) of the Building Regulations (Northern Ireland) 2000, see www.buildingcontrol-ni.com.

1.2.12 The Health and Safety (Safety Signs and Signals) Regulations 1996

These Regulations require employers to use a safety sign where there is a significant risk to health that cannot be controlled or avoided by other means. The HSE guide *Safety Signs and Signals* (L64) describes in detail a range of signs including Emergency Escape and Fire Safety Signs.

1.2.13 The Regulatory Reform (Fire Safety) Order 2005 (England and Wales)

This is probably the single most influential statutory instrument in the field of protection against fire. It has a direct influence on other pieces of primary and secondary statutory legislation, requiring modifications to and in some cases partial or full revocation of the requirements therein. It replaced fire certification under the Fire Precautions Act 1971 with a general duty to ensure, so far as is reasonably practicable, the safety of employees and a general duty, in relation to non-employees, to take such fire precautions as may reasonably be required to ensure that premises are safe. It also requires a risk assessment to be carried out and regularly updated.

The Regulatory Reform (Fire Safety) Order 2005 (henceforth referred to as 'the Order'), which came into effect fully in October 2006, affected over 70 pieces of fire safety law, not all of which fall within the scope of this Guidance Note.

The Order applies to non-domestic premises in England and Wales, including any communal areas of blocks of flats or houses in multiple occupation, and puts legal obligations and responsibilities upon the **responsible person** who is defined in article 3 of the Order as follows:

---

3. Meaning of 'responsible person'

In this Order 'responsible person' means –

(a) in relation to a workplace, the employer, if the workplace is to any extent under his control;

(b) in relation to any premises not falling within paragraph (a) –

(i) the person who has control of the premises (as occupier or otherwise) in connection with the carrying on by him of a trade, business or other undertaking (for profit or not); or

(ii) the owner, where the person in control of the premises does not have control in connection with the carrying on by that person of a trade, business or other undertaking.

Part 2 (Fire Safety Duties) of the Order requires the responsible person to take the necessary general fire precautions to ensure, so far as is reasonably practicable, the safety of employees and other persons within the premises for which they are responsible. The responsible person must carry out a fire safety risk assessment on the premises to identify what general fire precautions are required for the above to be achieved. The risk assessment must be regularly reviewed by the responsible person to keep it up to date.

The significant findings of the risk assessment, including any actions that have been taken or which will be taken by the responsible person and details of any persons identified as being especially at risk, must be recorded as soon as practicable afterwards where:

(a) the responsible person employs five or more persons, or

(b) the premises to which the assessment relates are subject to a licensing arrangement, or

(c) the premises are subject to an alterations notice.

Article 23 requires employees to:

(a) take reasonable care for the safety of themselves and of other persons who may be affected by their acts or omissions at work;

(b) cooperate with their employer as regards any duty or requirement imposed by or under any provision of the Order, so far as is necessary to enable that duty or requirement to be performed or complied with; and

(c) inform their employer or any other employee with specific responsibility for the safety of fellow employees –

    (i) of any work situation which they would reasonably consider represented a serious and immediate danger to safety; and

    (ii) of any matter which they would reasonably consider represented a shortcoming in the employer's protection arrangements for safety, in so far as that situation or matter either affects the safety of the employee or arises out of or in connection with their activities at work, and has not previously been reported to the employer.

GN2 Article 37 relates to firefighters' switches for luminous tube signs etc. and is a topic dealt with in Guidance Note 2: *Isolation & Switching*.

Reference should be made to the various schedules to the Order, which are as follows:

Schedule 1	This consists of four parts. Parts 1 and 2 relate to risk assessment, Part 3 relates to principles of prevention and Part 4 relates to measures to be taken in respect of dangerous substances.
Schedule 2	Amendments of primary legislation.
Schedule 3	Amendments of subordinate legislation.
Schedule 4	Repeals.
Schedule 5	Revocations.

It should be noted that in Scotland the Fire (Scotland) Act 2005, supported by the Fire Safety (Scotland) Regulations 2006 provides similar requirements for fire safety duties to those of the Regulatory Reform (Fire Safety) Order 2005 applicable to England and Wales and that in Northern Ireland the Fire Safety Regulations (Northern Ireland) 2010 apply.

1.2.14 Other Regulations

Electrical installations are subject to many local authority by-laws and conditions of licence. The following may be mentioned:

1 places licensed for public entertainment, music, dancing etc;

2 large buildings, office blocks and the like, having limited access and escape points; and

3 buildings to which the public have access, covered shopping centres and the like, hotels and boarding-houses.

Section 20 of the London Building Acts (Amendment) Act 1939 (as amended primarily by the Building (Inner London) Regulations 1985) is principally concerned with the danger arising from fire within certain classes of building which, by reason of height, cubic capacity and/or use, necessitate special consideration. The types of building coming within these categories are defined under Section 20 of the amended 1939 Act.

Note: These regulations have been subject to some repeals under the Regulatory Reform (Fire Safety) Order 2005. See Schedule 4 of the Order for full details.

The designer of the electrical installation in any building falling within these descriptions should make the earliest possible approach to the Local Authority and the Local Fire Authority to establish their requirements, obtain advice and get written agreement to the proposed arrangements.

Overview of the Wiring Regulations

2

This Guidance Note follows, as far as possible, the sequence in which matters relating to protection against fire appear within the 18th Edition. This approach has been adopted in order to make for its easy use in conjunction with the 18th Edition itself.

Chap 13 Chapter 13 of BS 7671 prescribes the fundamental principles for safety and includes a number of regulations which are directly concerned with the fire risk associated with electrical installations.

131.3.1
131.3.2 Regulation 131.3 sets out the general concept that the installation shall be designed and arranged to protect against the harmful effects of heat, arcs and burns.

Chapter 42, 'Protection against thermal effects', has requirements for:

421 **(a)** protection against fire caused by electrical equipment, including some particular requirements concerning the enclosures of consumer units and similar switchgear assemblies in domestic (household) premises;

422 **(b)** precautions where particular risks of fire exist, and

423 **(c)** protection against burns.

Sect 421 Section 421, 'Protection against fire caused by electrical equipment', requires measures to be taken to prevent electrical equipment from presenting a fire hazard to materials that are, or could conceivably be, found in close proximity to such equipment.

Sect 422 Section 422, 'Precautions where particular risks of fire exist', contains a number of specific requirements for wiring systems and switchgear/controlgear located within escape routes. Concern is often expressed over the impairment of escape routes by the smoke and fumes generated by fire. This is an important but very complex issue, and the risk from electrotechnical products should not be considered separately but as a part of the overall fire hazard assessment.

422.4 Section 422 also contains measures in addition to those of Section 421 to be applied in installations or locations therein in buildings mainly constructed of flammable materials such as wood.

Sect 423 Section 423, 'Protection against burns', states maximum temperature limits not to be exceeded on the external surfaces of fixed electrical equipment in normal use.

Chap 43 Chapter 43 prescribes requirements which limit the temperatures of live conductors and their insulation to prevent damage to the cable under overload and fault conditions.

522.1
Table 52.1
Table 43.1 Regulation 522.1.1 requires a wiring system to be installed such that it is suitable for both the highest and lowest ambient temperatures likely to be experienced. Further, it requires a cable to be selected and installed such that the limiting temperature of its insulation as stated in Table 52.1, and its limiting temperature under fault conditions given in Table 43.1, will not be exceeded.

Sect 527 Section 527 prescribes requirements for wiring systems to be installed to minimise the spread of fire, which are described in Chapter 6 of this Guidance Note.

Sect 532 Section 532 gives requirements for devices being selected to provide protection against fire.

Chap 54
GN6 Chapter 54 has the same intent as Chapter 43 as regards protective conductors under earth fault conditions. Both these chapters are the subject of Guidance Note 6 *Protection Against Overcurrent*.

Chap 55
554.4.4
559.4.1 Chapter 55 gives requirements for items of other equipment. Within this chapter, requirements pertinent to protection against fire are given for heating conductors and cables, and for low voltage luminaires and lighting installations.

Chap 56 Chapter 56 is of relevance to this Guidance Note, as it gives requirements for safety services.

Sect 642
Sect 642.3 Section 642 contains an inspection requirement for the presence and, indeed, effectiveness of fire barriers, seals and similar measures provided for protection against thermal effects.

651.2 Section 651 contains specific requirements for periodic inspection to provide for the safety of persons and livestock with regard to electric shock and burns, and protection against damage to property caused by fire and heat resulting from an installation defect.

705.422.7 Section 705, 'Agricultural and horticultural premises', includes a requirement for the installation of an RCD having a rated residual operating current not exceeding 300 mA for additional protection against fire in some circumstances.

Sect 711 Section 711 contains a number of requirements relating to the installation of luminaires and other items of equipment having high surface temperatures installed in installations for exhibitions, shows and stands.

Sect 715
715.42 Section 715 'Extra-low voltage lighting installations', applies to extra-low voltage lighting installations supplied from a source with a maximum rated voltage of 50 V AC or 120 V DC and includes a number of requirements relating to protection against thermal effects for such installations.

740.422.3.7 Section 740 requires a manually resettable thermal cut-out to be fitted to motors which are automatically and/or remotely controlled and unsupervised where installed in amusement parks, circuses, fairgrounds and similar temporary installations.

Sect 753
753.42 Section 753 deals with heating cable and embedded heating systems and contains a number of requirements relating to protection against burns and overheating, a

particular requirement being that the floor temperatures are limited to no more than 35 °C where contact with skin or footwear is possible.

For additional information reference can be made to CENELEC Guide 29 *Temperatures of Hot Surfaces Likely to be Touched*. The document provides guidance for assessing the risk of a burn from unintentional contact with readily accessible surfaces of electrical equipment operating with a voltage of between 50 V and 1000 V AC and 75 V and 1500 V DC. The document establishes surface temperature limits, where such limits are required, and describes the maximum contact periods with a hot surface that a person may be subjected to without being exposed to a risk of burn.

Where consideration of the effects of smoke, fumes, fire propagation and/or circuit integrity is necessary, cables which have characteristics classified in standard fire tests can be used. The cable standards are:

(a) BS 6724 – *Electric cables. Thermosetting insulated, armoured cables of rated voltages of 600/1000 V and 1900/3300 V, for fixed installations having low emission of smoke and corrosive gases when affected by fire specification.*
(b) BS 7211 – *Electric cables. Thermosetting insulated and thermoplastic sheathed cables for voltages up to and including 450/750 V, for electric power, lighting and internal wiring, and having low emission of smoke and corrosive gases when affected by fire*
(c) BS 7629-1 – *Electric cables. Specification for 300/500 V fire resistant, screened, fixed installation cables having low emission of smoke and corrosive gases when affected by fire. Multicore cables*
(d) BS 7846 – *Electric cables. Thermosetting insulated, armoured, fire-resistant cables of rated voltage 600/1000 V for fixed installations, having low emission of smoke and corrosive gases when affected by fire. Specification*

Cables clipped direct may also be mineral insulated to BS EN 60702-1. These may be either bare or, where a covering is required, it should be designated as having 'low emission of smoke and corrosive gases when affected by fire'.

General guidance on the selection of cables is given in BS EN 50565-1 – *Electric cables. Guide to use for cables with a rated voltage not exceeding 450/750 V (U_o/U).*

2

Generally applicable requirements | 3

3.1 General

131.1 The primary purpose of BS 7671 is to ensure that persons, property and livestock are protected against hazards arising from the electrical installation. The particular risks identified are:

(a) shock currents;

(b) excessive temperatures which may cause injuries, such as burns, and fire;

(c) the ignition of a potentially explosive atmosphere;

(d) undervoltages, overvoltages and electromagnetic influences which are potentially damaging or which could cause injury;

(e) injury from mechanical movement of electrically actuated equipment;

(f) power supply interruptions/interruption of safety services;

(g) arcing or burning that is likely to cause blinding effects, the production of excessive pressure and/or toxic gases and;

(h) arcing that is likely to cause fire.

Of the above, excessive temperatures, ignition of potentially explosive atmospheres, and arcing and burning effects are considered in this Guidance Note.

131.3.2 Persons, livestock, fixed equipment and fixed materials, whether part of the fabric of the building or the contents thereof, should be protected against harmful thermal effects which may result in combustion, ignition or degradation, the risk of burns, or the impairment of the safe functioning of equipment and hence the safe use of the installation for its intended purpose.

421.1 Harmful effects can occur as a result of:

(a) heat accumulation;

(b) radiated heat;

(c) hot components or items of equipment;

(d) failure of protective devices, switchgear, thermostats, temperature limiting devices;

(e) inadequate provision or failure of sealing arrangements where wiring systems penetrate the fabric of a building or within the wiring systems themselves (where required);

(f) overcurrent;

(g) insulation faults;

(h) arcing, sparking or high temperature particles;

(i) harmonic currents; and

(j) lightning.

3.2   Fixed equipment

Sect 421 Although there is a tendency for fires of unexplained cause to be blamed upon faulty electrical installations and electrical appliances, many fires can be correctly attributed to faults within, or incorrect selection and use of, either the fixed installation or electrical equipment. Neither the electrical installation designer nor the installer has any control over the selection and use of mobile (portable) equipment and so they are not considered within this Guidance Note. Chapter 42 is concerned solely with fixed equipment. Other sections of BS 7671 contain measures to minimise the risk of fire specific to particular types of equipment and these are summarised below.

3.2.1   Heating cables

554.4.1 Where it is necessary for a heating cable to either pass through or be in close proximity to any material which presents a fire hazard, it should be enclosed in material having the ignitability characteristic 'P' as specified in BS 476-12:1991 *Fire tests on building materials and structures. Method of test for ignitability of products by direct flame impingement.*

554.4.4 Furthermore, the loading of a floor-warming cable should be so limited that the manufacturer's maximum stated operating temperature is not exceeded in normal operation, due consideration being given to other factors which may further limit the maximum operating temperature, such as terminations, accessories and materials with which they are likely to be in contact.

Sect 753 Reference should also be made to Section 753, which contains requirements for heating cables and embedded heating systems.

3.2.2   Luminaires

559.4.1
422.3.1
422.4.2
The thermal effects of radiant and convected energy on the surroundings should always be considered when luminaires are selected and installed. Factors to be considered include: the fire resistance of materials in close proximity to, or likely to be affected by, the heat produced by the luminaire; confirmation that recommended minimum clearances between luminaires and combustible materials have been observed (see also requirements of Regulations 422.3.1 and 422.4.2, which are discussed further in Sections 4.3 and 4.4 respectively of this Guidance Note); and the use of lamps of the correct type and which do not exceed the maximum rating for the luminaire.

3.2.3   Extra-low voltage lighting installations

Sect 715
715.42
Extra-low voltage lighting installations are covered in Section 715 of BS 7671 Part 7. See Section 8.6 of this Guidance Note.

Like all special installations or locations, extra-low voltage lighting installations are still subject to the relevant general requirements of BS 7671, but these are supplemented and modified by the requirements of Section 715.

3.2.4   Symbols used in connection with luminaires, controlgear for luminaires and their installation

Table 55.3 The following symbols, from Table 55.3 of BS 7671, which may be found on luminaires and controlgear for luminaires or their packaging data sheets, are of relevance to the subject of protection against fire.

Table 3.1 Explanation of symbols used in luminaires, in controlgear for luminaires and in the installation of luminaires (Table 55.3 of BS 7671)

BS EN 60598-1:2008

 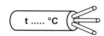
Recessed luminaire not suitable for direct mounting on normally flammable surfaces

Surface mounted luminaire not suitable for direct mounting on normally flammable surfaces

Luminaire not suitable for covering with thermally insulating material

Use of heat-resistant supply cables, interconnecting cables, or external wiring (number of conductors of cable is optional)

(BS EN 60598 series)

Luminaire for use with high pressure sodium lamps that require an external ignitor

(BS EN 60598 series)

Luminaire designed for use with bowl mirror lamps

(BS EN 60598 series)

Luminaire for use with high pressure sodium lamps having an internal starting device

(BS EN 60598 series)

ta .... °C
Rated maximum ambient temperature

(BS EN 60598 series)

Luminaire with limited surface temperature

(BS EN 60598-2-24)

Warning against the use of cool-beam lamps

(BS EN 60598 series)

Short-circuit proof (inherently or non-inherently) safety isolating transformer

(BS EN 61558-2-6)

Minimum distance from lighted objects (metres)

(BS EN 60598 series)

Temperature declared thermally protected lamp controlgear (... replaced by temperature)

(BS EN 61347-1)

Rough service luminaire

(BS EN 60598 series)

Electronic convertor for an extra-low voltage lighting installation

BS EN 60598-1:2008

Replace any cracked protective screen

(BS EN 60598 series)

Thermally protected lamp controlgear (class P)

(BS EN 61347-1)

Luminaire designed for use with self-shielded tungsten halogen lamps or self-shielded metal halide lamps only

(BS EN 60598 series)

Independent ballast

BS EN 60417 sheet No. 5138

Note: As required by BS EN 60598-1 2004 luminaires suitable for direct mounting on normally flammable surfaces may be marked with the symbol shown (triangle with F inside) according to BS EN 60598-1:2004.

With the publication of BS EN 60598-1 2008, luminaires suitable for direct mounting on normally flammable surfaces have no special marking and only luminaires not suitable for mounting on normally flammable surfaces are marked with a symbol (see Annex N.4 of BS EN 60598-1:2008 for further guidance). Products may therefore be sold with the old and/or new symbols displayed on them or their packaging.

3.3   Manufacturers' instructions

421.1
134.1.1
510.3
When electrical equipment is installed, any reasonable and relevant instructions of the equipment manufacturer should be taken into account. This is a reminder of the general requirement that electrical equipment be installed in accordance with any instructions provided by the manufacturer. Where an installation includes fixed equipment, the fact that the selection of the equipment may have been made by someone other than the installer does not absolve the installer from the responsibility of seeing that any installation instructions of the manufacturer are met.

3.4   No fire risk in normal use

421.1.2
The heat generated by fixed electrical equipment in normal use should not be capable of causing a fire or harmful thermal effects to adjacent fixed material. Further, the installation designer is required to exercise foresight and consider other material which is likely to be in proximity to such equipment.

Regulation 421.1.2 offers three installation methods for equipment which in normal operation has a surface temperature sufficient to cause a risk of fire or harmful effects to adjacent materials. In short, such equipment should be supported, enclosed or screened by items capable of withstanding the operating temperature, or be mounted in a manner that permits safe heat dissipation and gives adequate clearance from surrounding equipment or materials.

One or more of the methods must be adopted and as far as Regulation 421.1.2(iii) is concerned, Figure 3.1 gives general guidance for equipment which in normal operation has a surface temperature exceeding 90 °C.

▼ **Figure 3.1** Clearances required to achieve requirements of Regulation 421.1.2(iii)

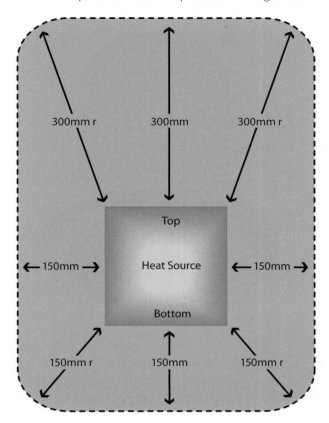

By way of example, attention must be given to recessed or semi-recessed luminaires and luminaire controlgear mounted in ceiling voids to ensure that heat is properly dissipated and that any thermally insulating materials are not and cannot become so disposed as to restrict the cooling of the equipment.

This might be achieved by the use of a fully closed back 'fire-rated' type downlighter such as the one shown in Figure 3.2, which is particularly suitable for installation into ceiling/floor voids where access above the luminaire is not possible.

▼ **Figure 3.2** Fire-rated downlighter

Where access to the space above the luminaire can be achieved, such as for example in a domestic loftspace, a proprietary insulation support box such as the one shown in Figure 3.3 may suffice if installed over more open-backed design recessed luminaires.

▼ **Figure 3.3** Thermal insulation support box

559.5.1.205 Lampholders are one example of equipment having temperature ratings. BS EN 61184:2008 *Bayonet lampholders*, gives guidance on the selection of lampholders for particular applications. However, because it is not possible to guarantee how an installation will be used, every B15 or B22 lampholder must be T2 temperature rated to comply with BS 7671. Where lampholders other than B15 or B22 are installed, their rated operating temperature must be suitable for the application.

3.5 Arcs, sparks and hot particles

421.1.3 The effects of arcs or the emission of high temperature particles must be guarded against. In some equipment, for example air circuit-breakers or semi-enclosed fuses, emissions may be produced during fault clearance and the enclosures of such equipment must comply with the appropriate British Standard, be totally enclosed in arc-resistant material, or be screened by arc-resistant material, or be mounted so as to allow for the safe extinction of the emissions at a sufficient distance from the material upon which the emissions could have a harmful effect.

Note that electric arc welding and similar equipment (fixed or mobile) are subject to Regulations 7, 8 and 14 of the Electricity at Work Regulations 1989.

3.6 Protection against arcing

421.1.7
532.6 Additional protection to mitigate against the risk of fire due to arcing is recommended. Regulation 421.1.7 recommends the use of arc fault detection devices (AFDD) conforming to BS EN 62606 as a means of providing additional protection against fire caused by arc faults in AC final circuits.

If used, an AFDD shall be placed at the origin of the circuit to be protected.

Examples of where such devices can be used include:

(a) premises with sleeping accommodation;
(b) locations with a risk of fire due to the nature of processed or stored materials, i.e. BE2 locations (e.g. barns, wood-working shops, stores of combustible materials);
(c) locations with combustible constructional materials, i.e. CA2 locations (e.g. wooden buildings);
(d) fire propagating structures, i.e. CB2 locations; and
(e) locations with irreplaceable objects (e.g. museums, libraries, art galleries etc.).

AFDDs conforming to BS EN 62606 protect against series and parallel arcing and are designed to detect low level hazardous arcing that circuit breakers, fuses and RCDs are not designed to detect. Such arcing could occur due to, for example, aged/damaged/pierced cable insulation, mechanically damaged/stressed cables, loose (arcing) terminations. They are intended to mitigate the effects of arcing faults by disconnecting the circuit when an arc fault is detected, thus preventing the possible outbreak of fire. They detect low level hazardous arcing that circuit breakers, fuses and RCDs are not designed to detect.

AFDDs are designed and tested to not respond to arcing under normal operation of equipment such as vacuum cleaners, drills, dimmers, switch mode power supplies, fluorescent lamps, etc., in addition, they are designed and tested to continue to respond to arcing faults whilst such equipment is being operated.

3.7   Focusing and concentration of heat

421.1.4   Equipment that focuses heat, such as radiant heaters and high intensity luminaires, must be positioned so that the building structure or other materials are not subjected to excessive temperatures.

422.3.1   The manufacturers' instructions must be followed with respect to spacing from walls (w), floor (f) and ceiling (c), and the angle of inclination q of the heater (Figure 3.4). Regulation 422.3.1 states the **minimum** distances that spotlights and projectors must be installed from combustible materials; see Section 4.3.1 of this Guidance Note.

Downlighters fitted with aluminised lamps project both the light and heat generated away from the light fitting. This heat can be considerable, therefore care should be taken to ensure that luminaires fitted with such lamps are not so placed that, for example, a door can be left open below sufficiently close for a risk of fire to exist.

See also Section 3.2 of this Guidance Note.

▼ **Figure 3.4** Focusing of heat from a heater or luminaire

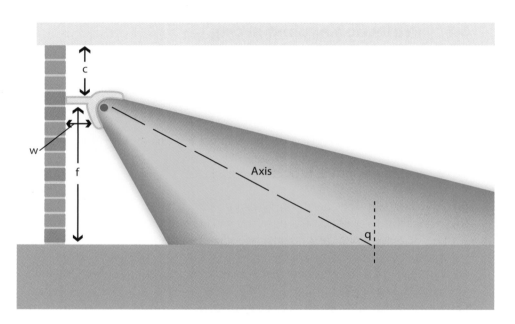

3.8   Equipment containing flammable liquid

421.1.5   Special precautions are necessary for flammable dielectric liquids, as fires involving these liquids are life-threatening within a few seconds of ignition. In the event of spillage, the object is to limit the spread of the liquid and the exposed surface area, thus limiting the size of the fire and the danger to persons and property.

Where a large quantity of liquid is involved, more than one escape route is necessary, in the same way as for a large switchroom. It is important to ensure that persons have alternative routes so that the main seat of a fire may be avoided.

The 'single location' referred to in Regulation 421.1.5 is the location containing all the flammable liquid which may be involved at the outset of an emergency. The amount of flammable liquid in an area must therefore be established, whether contained in one item of equipment or in a number of separate items.

The options available to the designer will depend on a number of considerations, for example, whether a single item of equipment is involved or a number of items and whether the location is indoors or outdoors. The options include:

(a) reducing the risk by partitioning the location with fire doors and sills;

(b) providing bunds or kerbs around items of equipment or, for larger items of equipment, a retention pit filled with pebbles or granite chips (the net capacity of the bund or retention pit when filled with pebbles or chips should exceed the oil capacity of the equipment by at least 10 %);

(c) providing a drainpit and flame arrestor;

(d) provision of automatic fire venting and/or automatic fire suppression or foam inlets and integration with the automatic fire detection and alarm system of the building, where appropriate;

(e) ramped floors;

(f) use of an outdoor location; and

(g) having blast walls between large items.

▼**Figure 3.5** An oil-filled transformer located in a plant room, which incorporates many of the measures described above: the location is fitted with a fire door, the floor falls to a central drainage point leading to a drainpit (grille just visible to left of picture) and the walls have been constructed to contain any blast in the event of explosion

In some circumstances, it may be appropriate to provide explosion venting for interior locations, and environmental considerations should always involve discussions with public health engineers concerning the provision of an oil interceptor to prevent contamination of the sewers.

When carrying out inspections of older installations the measures provided should be carefully examined in the light of modern practice. In particular, it may be that over a period of time, seepage and occasional spillage has caused a wooden floor to be saturated with oil and thus become an additional fire risk.

Adequate general lighting and emergency lighting should be provided where appropriate (particularly on escape routes).

First-aid equipment should be provided where appropriate.

3.9 Enclosures

421.1.6 Where enclosures are constructed on site, reference must be made to the appropriate product standard for the necessary resistance to heat and fire. In the absence of such a standard, maximum likely temperatures of the enclosure must be determined and enclosures selected that are well able to withstand these temperatures.

BS 476 *Fire tests on building materials and structures (Part 4 – Non-combustibility test for materials, Part 12 – Method of test for ignitability of products by direct flame impingement)* provides type tests for materials.

Where equipment is installed in an enclosure or enclosed space, for example ballasts or other controlgear within a ceiling, adequate arrangements must exist for safely dissipating heat generated.

3.9.1 Consumer units and similar assemblies in domestic premises

421.1.201
132.12 Regulation 421.1.201, requires consumer units and similar switchgear assemblies installed in domestic (household) premises to comply with BS EN 61439-3 and to either:

(a) have their enclosure manufactured from non-combustible material or
(b) be enclosed in a cabinet or enclosure constructed of non-combustible material and complying with Regulation 132.12.

An example of a non-combustible material, according to Note 1 to Regulation 421.1.201, is ferrous metal, such as steel.

Where option (b) above is used, it is important that the cabinet or enclosure is suitably selected and erected to meet the requirements of Regulation 132.12. These relate to adequacy of space for initial installation and later replacement individual items equipment, and accessibility of the equipment for operation, inspection, testing, fault detection, maintenance and repair.

The intent of Regulation 421.1.201 is, as far as is reasonably practicable, to contain any fire within the non-combustible enclosure or cabinet and to minimise the emission of flames to the surroundings or into conduits trunking or ducting. Consequently, both of the following are necessary, irrespective of which of options (a) or (b) is adopted.

(a) The non-combustible enclosure or cabinet must provide a complete envelope (e.g. base, cover, door and any components such as hinges, screws and catches) as necessary to maintain fire containment. All blanks, circuit-breakers and other devices must be contained within the non-combustible enclosure or cabinet.
(b) The installer must seal all openings into the non-combustible enclosure or cabinet for cables, conduits, trunking or ducting that remain after the installation of cables. Good workmanship and proper materials must be used and account must be taken of the manufacturer's relevant instructions, if any.

The phase "similar switchgear assemblies" in Regulation 421.1.201 means assemblies used for the same fundamental application as a consumer unit. An example could be a three-phase distribution board intended to be operated by ordinary persons.

Regulation 421.1.201 is intended to apply to consumer units and similar switchgear assemblies to BS EN 61439-3 inside all domestic (household) premises, including their integral/attached garages and outbuildings and those in close proximity.

3.10 Terminations

526.5 Regulation 526.5 is the general requirement for the enclosure of terminations and joints of live conductors and PEN (combined protective and neutral) conductors. A further important requirement is to provide an enclosure which will protect the joint or connection against the environment.

**526.1 to
526.9
642.3** Poor terminations and connections are cited as a frequent cause of fire, and close attention is required by the designer, the installer and the person responsible for the inspection and testing to all aspects of the subject.

For this reason, Regulation 526.5 requires that all terminations and joints at whatever voltage, including ELV, shall be made within a suitable enclosure. Because there is always a risk of overheating and consequent fire at a joint or termination, these must be enclosed and the enclosure where not incorporated in suitable equipment or accessory must meet the specified fire resistance requirements. This applies equally to LV and ELV connections, to luminaires and to similar equipment. The relatively high currents of ELV equipment mean particular care has to be taken with joints and terminations. When compression joints are used, crimping tool, lug and cable must be compatible.

Other matters which should be considered under this heading are:

(a) dirty or misaligned equipment contacts which may give rise to local heating; and
(b) loose or inadequate cable supports which may place mechanical stresses on connections, causing overheating.

One method, which could be considered when inspecting electrical installations for signs of overheating, is the use of thermal imaging. This is not required by BS 7671 and would therefore be in addition to the inspections and tests stated in Part 6 of BS 7671.

The use of thermographic surveying is discussed in Guidance Note 3: *Inspection & Testing.*

3.11 Installation of Cables

521.10.202 Regulation 521.10.202, requires wiring systems to be supported in such a way that they will not be liable to premature collapse in the event of fire.

The purpose of this regulation is prevent wiring systems supports failing under fire conditions and cables dropping or hanging across access or egress routes which may have the potential to hinder evacuation and firefighting activities.

The requirements of Regulation 521.10.202 preclude the use of non-metallic cable clips or cable ties as the sole means of support where cables are clipped direct to exposed surfaces or suspended under cable tray. The use of non-metallic cable trunking is also precluded as the sole means of support of the cables. Suitably spaced steel or copper clips, saddles or ties are examples that will meet the requirements of this regulation. The cables must be supported at appropriate intervals by suitable fire-resistant means, securely fixed, that will prevent the cables from falling.

Further guidance on methods of support for cables, conductors and wiring systems can be found in the Appendix D of the *On-Site Guide*.

Further requirements relating to wiring systems in escape routes apply for:

(a) locations where particular risks of fire exist (see Chapter 4 of this Guidance Note); and
(b) safety services (see Chapter 7 of this Guidance Note).

Precautions where particular risks of fire exist 4

4.1 General

Sect 422 BS 7671 considers particular risks of fire to exist in escape routes from a building; locations having risks of fire due to the nature of materials being processed or stored therein; where the fabric of the building has been constructed mainly of combustible materials; where the structure of the building aids the propagation of fire; and where the building itself or indeed its contents are of some particular significance.

422.1.1 Generally, in such locations, only equipment necessary for the intended use of the location should be installed. Exceptions to this requirement are given in Regulation 422.3.5 and are considered later.

Any installed equipment should be so constructed that any normal or foreseeable temperature rise, even if present under fault conditions, cannot result in a fire occurring. If necessary this requirement may be met by the application of additional protective measures.

422.1.3 Any thermal cut-out devices installed in a location covered by Section 422 should have a manual reset facility.

4.2 Requirements for emergency escape routes

Appx 5 Those parts of a building which constitute part of an emergency escape route are
422.2 classified in Appendix 5 of BS 7671. Regulation 422.2 makes it clear that no special requirements apply in conditions having the classification BD1 (low density/easy access), which is domestic premises and other places of low density occupation and of low to normal building height.

Regulation 422.2 does contain a number of specific requirements for locations classified as BD2 (multi-storey buildings such as offices, for example), BD3 (buildings open to the public, such as shopping centres and places of public entertainment, for example) and BD4 (high-rise buildings open to the public, such as hotels, for example).

422.2.1 Wiring systems should not encroach on an escape route and should in any case be as short as practical. Further, only wiring systems having sufficient protection against mechanical damage likely to be encountered during evacuation in such a location may be installed within arm's reach.

The means of support for wiring systems must meet the requirements of Regulation 521.10.202 to prevent premature collapse of the wiring system in the event of fire. Section 3.11 of this Guidance Note refers.

In addition, in the emergency escape routes of locations covered by this chapter, all wiring systems employed, including both the cables themselves and the supporting structure provided, must be non-flame propagating.

This can be achieved by the use of:

(a) cables meeting the appropriate fire tests prescribed in BS EN 60332-3 and the requirements of BS EN 61034-2;

(b) conduit systems meeting the fire tests prescribed in BS EN 61386-1;

(c) trunking and ducting systems meeting the fire tests prescribed in BS EN 50085;

(d) cable tray and cable ladder systems to BS EN 61537; and

(e) powertrack to BS EN 61534.

Any wiring within the escape route which supplies a safety service must, in the absence of a more stringent requirement for building elements, have a resistance to fire of one hour.

A decent level of visibility in escape routes should be maintained for as long as possible. To achieve this, wiring in such locations should be of a type having limited smoke production characteristics. Where no more stringent requirement exists in a cable product standard, a value of 60 % light transmittance is recommended for cables tested to BS EN 61034-2.

422.2.2
Part 2
Any switchgear or controlgear other than fire alarm call points installed in a building classified BD2, BD3 or BD4 should be so installed that it is only accessible to authorised persons. Further, if such items are installed within an emergency escape route an additional enclosure made of non-combustible or not readily combustible material should be provided. This requirement does not extend to accessories such as light switches (see Definitions).

422.2.3
With the exception of small capacitors likely to be found in equipment such as discharge lights, equipment containing flammable liquids should not be installed in escape routes.

Reference should be made to the relevant parts of the BS 5266 series, which gives the requirements for emergency escape lighting.

4.3 Risks of fire due to the nature of processed or stored materials

Appx 5
Appendix 5 of BS 7671 classifies the risks associated with the nature of processed or stored materials. Classifications relevant to protection against fire are:

(a) BE2 – a fire risk exists as a direct result of the manufacturing, processing or storage of materials. This would include premises such as barns, woodworking shops, industrial scale bakeries and paper mills.

(b) BE3 – an explosion risk exists from the processing or storage of low flashpoint materials. This would include petrochemical plants and hydrocarbon fuel storage facilities.

BS 7671 contains no specific requirements where an explosion risk exists, but rather refers to BS EN 60079-14 *Explosive atmospheres. Electrical installation design, selection and erection.*

Regulation 422.3 contains a number of requirements (described below) to be applied in addition to the general requirements for installations in locations classified as BE2, where the manufacturing, processing or storage of combustible materials and the accumulation of materials such as dust and fibres constitutes a risk of fire but not explosion.

4.3.1 Luminaires

422.3.1 Except where otherwise recommended by the manufacturer, spotlights and projectors should be installed at the following minimum distances from combustible materials:

(a) rating up to 100 W – 0.5 m
(b) rating over 100 W up to 300 W – 0.8 m
(c) rating over 300 W up to 500 W – 1.0 m.

422.3.8 Only luminaires having enclosures providing a degree of protection of at least IP4X, or IP5X or IP6X in the presence of dust or electrically conductive dust, respectively, and with limited surface temperatures in accordance with BS EN 60598-2-24, should be employed. Fittings manufactured to BS EN 60598-2-24 have horizontal surface temperatures limited to 90 °C under normal conditions and 115 °C under fault conditions; these will be marked ▽D̲.

Additionally, luminaires should be of a type that prevents components which are likely to run hot, such as lamps, from falling from the luminaires.

4.3.2 Enclosures

422.3.2 Enclosures of equipment such as heaters should not be able to attain higher surface temperatures than 90 °C under normal conditions, or 115 °C under fault conditions.

Where it is likely that dust or fibres could accumulate on an enclosure of electrical equipment in such a manner or quantity as to cause a fire hazard, precautions need to be taken to prevent said enclosure from exceeding the aforementioned temperatures. Good housekeeping is also necessary in such locations.

4.3.3 Switchgear and controlgear

422.3.3
GN1 Appx B Electrical equipment shall be installed outside the location unless suitable for the location, or installed inside a suitable enclosure. A suitable degree of protection for such an enclosure would be IP4X, or IP5X or IP6X if the presence of dust or conductive dust, respectively, is to be reasonably expected. The IP code system is covered by BS EN 60529 *Specification for degrees of protection provided by enclosures (IP code)*. To summarise, the degree of protection provided by an enclosure is indicated by two numerals. The first indicates the degree of protection against solid bodies and the second indicates the degree of protection against liquids. Where a characteristic numeral is not required to be specified it can be replaced by the letter X (refer to Guidance Note 1, Appendix B for full details). For IP4X, the first numeral 4 indicates protection against contact by wires or strips more than 1.0 mm thick. For IP5X, the first numeral 5 indicates dust-protected (dust may enter but not in an amount sufficient to interfere with satisfactory operation or impair safety). For IP6X, the first numeral 6 indicates dust-tight (no ingress of dust).

4.3.4 Cables, circuits and wiring systems

422.3.4 Cables not completely embedded in non-combustible material, such as plaster or concrete, or otherwise protected from fire shall meet the requirements of BS EN 60332-1-2.

Conduit systems should be able to satisfy the test under fire conditions specified in BS EN 61386-1. Ducting and trunking systems should be able to satisfy the test under fire conditions specified in BS EN 50085.

Cable tray or cable ladder systems should be able to satisfy the test under fire conditions specified in BS EN 61537.

A powertrack system should be able satisfy the test for resistance to flame propagation specified in the appropriate part of the BS EN 61534 series.

Wiring systems shall be selected and installed to minimize the propagation of flame and where the risk of flame propagation is high the cable shall meet the requirements of the appropriate part of BS EN 60332-3 series.

The risk of flame propagation can be high where cables are bunched or installed in long vertical runs and where this risk exists the cable shall also meet the requirements of the appropriate part of BS EN 60332-3 series.

Cables manufactured for the above application also need to satisfy the requirements of the Construction Product Regulations in respect of their reaction to fire and manufacturers must declare the performance of the cables in terms of their reaction to fire properties as contained in BS EN 50575. The Regulations apply to all cables for use in fixed installations in domestic, commercial and industrial premises and other civil engineering works. The requirement applies to power, communications and fibre optic cables. Testing and classification need to be carried out by an independent Notified Body who will issue a Declaration of Performance (DoP).

Designers will be required to reference the cables that are to be installed in projects and installers will need to purchase and use products that are complaint with the design and provide an audit trail to show that compliant cables were installed.

422.3.5
422.3.10 A wiring system which passes through, but is not intended for electrical supply within, the location containing combustible stored material, should meet the requirements of Regulation 422.3.4 above, and have no joint or connection within the location unless it is placed in a suitable enclosure that does not adversely affect the flame propagation characteristics of the wiring system. Circuits originating inside such locations should be suitably protected against overcurrent at their origin.

422.3.9 With the exceptions of mineral insulated cables complying with BS EN 60702-1, powertrack systems complying with BS EN 61534 and busbar trunking systems complying with BS EN 61439-6, wiring systems should be protected against insulation faults to earth as follows:

(a) In TN and TT systems, by an RCD with a rated residual operating current not exceeding 300 mA. Where a resistive fault in, for example, an overhead heating system employing heating film elements may cause a fire, 30 mA RCD protection will be required.

(b) In IT systems, by insulation monitoring devices with audible and visual signals. Adequate supervision is required to facilitate manual disconnection as soon as appropriate. In the event of a second fault, the disconnection time should be in accordance with the requirements of Regulation 411.6.4.

532.1　In locations where, in accordance with Chapter 42, a particular risk of fire exists, preventive protection measures against the risk of fire will be required. A risk analysis will determine whether these protective measures will need to be applied to other parts of the electrical installation.

532.1
532.2　Where an RCD is used to meet the requirements of Regulation 422.3.9 as described above, the RCD should:

(a) be installed at the origin of the circuit for which it provides protection; and
(b) have a rating not exceeding 300 mA.

422.3.11　All live parts of extra-low voltage circuits, regardless of the nominal voltage, must either:

1 be contained in an enclosure having a degree of protection of at least IP2X or IPXXB, or
2 have insulation capable of withstanding a test voltage of 500 V DC for 1 minute.

422.3.12
543.4.1
543.4.2　The use of PEN conductors is not permitted for circuits within the location, but is permitted for wiring passing through the location. It should be remembered that the Electricity Safety, Quality and Continuity Regulations 2002 prohibit the use of PEN conductors in consumers' installations, with very few exceptions which are described in Regulation 543.4.2.

PEN conductors are most likely to form part of either a public or private supply arrangement.

422.3.13
537.1.2　Every circuit, subject to the exceptions given in Regulation 537.1.2, should be provided with a means of isolation which disconnects all live conductors by means of a linked switch or linked circuit-breaker. If an installation or item of equipment or an enclosure contains live parts connected to more than one supply, a warning notice should be placed highlighting the need to isolate those parts from the various supplies unless an interlocking arrangement is provided.

4.3.5 Motors

422.3.7　All motors which are automatically or remotely controlled, or which are not continuously supervised, should be protected against overtemperature by a protective device with manual reset. Motors with star-delta starting must be protected against overtemperature when connected in all operating modes.

It is also advisable to protect motors with slip-ring starters from being left with the resistance in the rotor. This type of rotor has a 3-phase winding, the ends of which are connected to three slip-rings on the rotor shaft. For this reason it is sometimes called a 'slip-ring' motor. This enables an external resistance to be added to the rotor circuit, which is used to:

(a) limit the starting current;
(b) give a high starting torque; and
(c) provide speed control.

At normal running speed the slip-rings are short-circuited and the brushgear is lifted clear of the slip-rings to reduce wear.

4.3.6 Heating appliances

422.3.202
422.3.203
Any heating appliances installed in the location should be installed as fixed equipment, and heat storage appliances should be of a type which prevents the ignition of combustible materials and/or fibres by the heat storing core.

4.4 Combustible constructional materials

422.4
Regulation 422.4 has the following requirements where a building is mainly constructed of combustible materials such as wood (classified in Appendix 5 as CA2) and there is no explosion risk.

4.4.1 Equipment

422.4.1
Precautions should be taken so that electrical equipment does not ignite walls, floors or ceilings to which it is in close proximity, by the adoption of appropriate design, installation methods and choice of electrical equipment.

422.4.201
Distribution boards and accessory boxes for switches, socket-outlets and the like that are installed into or on the surface of a wall made from combustible materials should meet the requirements of the relevant product standard for temperature rise of such an enclosure and where recessed shall be at least IP3X.

422.4.202
Where this is not the case, the equipment or accessory should be enclosed by non-flammable material of suitable thickness, taking into account the nature of the material being employed.

4.4.2 Luminaires

422.4.2
Except where an appropriate product standard states specific requirements for spacings from combustible materials, luminaires of the spotlight or projector type should be installed at the following minimum distances from combustible materials used to form the fabric of the building:

(a) rating up to 100 W – 0.5 m
(b) rating over 100 W up to 300 W – 0.8 m
(c) rating over 300 W up to 500 W – 1.0 m.

Luminaires should be of a construction that prevents components that are likely to run hot, such as lamps, from falling from the luminaires.

A luminaire that is marked ▽F̵ in accordance with BS EN 60598-1:2004 was deemed suitable for installation directly on a surface classified as normally flammable.

However, with the publication of BS EN 60598-1:2008, luminaires suitable for direct mounting on normally flammable surfaces have no special marking and only luminaires not suitable for mounting on normally flammable surfaces are marked with a symbol ⌸ ⌸ . Reference should be made to Section 3.2.4 of this Guidance Note and Table 55.3 of BS 7671 for further guidance on symbols used in luminaires.

4.4.3 Cables and wiring systems

422.4.203 Any cables installed in premises mainly constructed of combustible materials should meet the requirements of 422.2.1

422.4.204 Conduit systems should be in accordance with and meet the fire resistance requirements of BS EN 61386-1, *Conduit systems for cable management general requirements.*

Similarly, trunking systems should be in accordance with and meet the fire resistance requirements of BS EN 50085-1, *Cable trunking systems and cable ducting systems for electrical installations. General requirements.*

4.5 Fire propagating structures

Appx 5 Buildings the shape, dimensions and layout of which facilitate the spread of fire are classified in BS 7671 Appendix 5 as CB2. Examples of such buildings would include high-rise buildings and buildings containing forced ventilation systems where a 'chimney effect' may exist having the detrimental effect of assisting a fire to remain burning.

422.5 The requirements of Regulation 422.5 need to be applied in addition to those of Section 421 in locations where CB2 conditions exist and in particular where the risk of flame propagation is high the cable shall meet the cable requirements of Regulation 422.2.1 (see BS 7671 Appendix 5 for the classification and codification of external influences).

A note to Regulation 422.5 points out that fire detectors may be provided in such structures, which could be used to activate measures such as the closing of fireproof shutters in ducts, troughs and trunking.

A further note to the regulation points out that boxes and enclosures according to BS EN 60670-1 and bearing the symbol 'H' may be used, for hollow walls, and that cables in accordance to BS EN 60332-3 may be used.

422.5.1 Regulation 422.5.1 requires steps to be taken so that the electrical installation in such buildings does not contribute to the spread of a fire.

Sect 527
527.2.1.2 Particular attention should be paid to reinstatement of structural integrity in relation to fire resistance where wiring systems penetrate the fabric of the building, and to the installation of fire-stopping materials within wiring systems where required by BS 7671.

Reference should also be made to Section 6.2 of this Guidance Note which discusses the sealing of wiring system penetrations.

4.6 Locations of national, commercial, industrial or public significance

BS 7671:2018 provides guidance for the selection and erection of electrical installations in locations of national, commercial, industrial or public significance. This would include museums, national monuments, airports, railway stations, laboratories, computer and data storage centres, and archiving facilities.

422.6 Regulation 422.6 requires compliance with Regulation 422.1 and consideration of the following:

(a) Installation of mineral insulated cables according to BS EN 60702

(b) Installation of cables with improved fire-resisting characteristics in case of a fire (i.e. shall meet the fire-resisting requirements of BS 7629-1, BS 7846 or BS 8573, as appropriate)

(c) Installation of cables in non-combustible solid walls, ceilings and floors

(d) Installation of cables in areas with constructional partitions having a fire-resisting capability for a time of 30 minutes. In locations housing staircases and required for emergency escape purposes a 90-minute fire-resistance capability is stated

(e) Where (a) to (d) are not practicable, improved fire protection may be possible by the use of reactive fire protection systems (e.g. by a holistic fire engineering approach to the location).

Protection against burns

5

423.1 The requirements of Regulation 423.1 of BS 7671 apply only to protection against burns caused by contact with heated surfaces. Measures to prevent burns from heat radiation or arcing are covered by the requirements of Regulation 421.1.1.

Table 42.1 Table 42.1 of BS 7671 (reproduced here as Table 5.1) gives maximum acceptable surface temperatures for accessible parts of equipment within arm's reach during normal load conditions.

▼ **Table 5.1** Temperature limit under normal load conditions for an accessible part of equipment within arm's reach

Accessible part	Material of accessible surfaces	Maximum Temperature (°C)
A hand-held part	Metallic	55
	Non-metallic	65
A part intended to be touched but not hand-held	Metallic	70
	Non-metallic	80
A part which need not be touched for normal operation	Metallic	80
	Non-metallic	90

Factors to be taken into account when using the table are whether the part is intended to be hand-held or touched in normal use, and of what materials the equipment is manufactured. Where the maximum temperatures prescribed in Table 42.1 are likely to be exceeded, albeit for a short period of time, the equipment in question should be fitted with guards or similar to prevent accidental contact.

Table 42.1 should not be applied to equipment for which a limiting temperature is specified in the relevant product standard.

It should be noted that mineral insulated cables exposed to touch are permitted to have a sheath temperature of 70 °C, corresponding to a metallic part intended to be touched but not hand-held. However, a cable having a conductor operating temperature of 90 °C may achieve a surface temperature approaching 80 °C in normal operation.

It must always be borne in mind that the temperatures in Table 42.1 are maximum values and that contact with any surface at or above 70 °C may cause a dangerous reflex action.

BS EN ISO 13732-1:2008, *Ergonomics of the thermal environment. Methods of the assessment of human responses to contact with surfaces. surfaces* provides information prepared by medical experts on human reaction to contact with heated surfaces. The document provides good guidance in determining 'safe' surface temperatures.

Even if the equipment complies with its standard as to surface temperature, consideration must still be given to the risk of burns, particularly when equipment is to be installed in locations to be used by the very young or infirm, where additional precautions may be necessary, such as guards over heaters.

554.2.1 Regulation 554.2.1 requires every heater of liquid or other substance to incorporate, or to be provided with, an automatic device to prevent a dangerous temperature rise of the substance being heated. This requirement would apply, for example, to immersion heater elements heating the water stored in cylinders for domestic premises.

Measures to minimise the spread of fire

6

6.1 Precautions in a fire-segregated compartment

Sect 527 Section 527 contains requirements regarding the selection of appropriate materials and the erection of the installation specifically aimed at minimising the spread of fire.

527.1.1 A fire-segregated compartment (fire compartment) is considered to be an enclosed space, which may be subdivided, separated from adjoining spaces within a building by elements of construction having a specified fire resistance.

527.1.1
527.1.2 In all cases, a wiring system should be installed such that it does not detrimentally affect the structural integrity or fire safety of the building.

527.1.3
527.1.5 In installations where no particular extra risks of fire exist, standard thermoplastic insulated cables and standard flexible cables that comply with the requirements of BS EN 60332-1-2 may be installed without the need for any other precautionary measures. However, where the fire-segregated compartment provides a means of evacuation in an emergency then the cable shall meet the requirements of the appropriate part of BS EN 60332-3 series.

Cables manufactured for the above application also need to satisfy the requirements of the Construction Products Regulation (CPR) in respect of their reaction to fire. Reference should be made to BS 7671 Appendix 2, item 17 for further guidance on manufacturers obligations in respect of testing, certifying, products intended for permanent installation in buildings and construction works.

Cables having the necessary resistance to flame propagation as specified in the BS EN 61386 series, the appropriate part of BS EN 50085 series, BS EN 61439-6, BS EN 61534 series, BS EN 61537 or BS EN 60570 may be installed without special precautions. Other cables that comply with standards having similar requirements for resistance to flame propagation may be installed without special precautions.

527.1.6
527.1.4 Alternatively, other cables can be installed in suitable non-combustible building materials.

The use of cables not complying with the cable requirements of BS EN 60332-1-2 (see Appendix A of this Guidance Note), should be limited to only short lengths making the connection between appliances and the fixed installation. Such cables should not be installed such that they pass from one fire segregated compartment to another.

Elements of a wiring system other than cables should meet the flame propagation requirements of the appropriate product standard as shown below:

(a) trunking and ducting – BS EN 50085
(b) conduit – BS EN 61386-1
(c) busbar trunking – BS EN 61439-6
(d) powertrack – BS EN 61534
(e) cable tray – BS EN 61537
(f) track systems for luminaires – BS EN 60570.

6.2 Sealing of wiring system penetrations

6.2.1 External sealing arrangements

527.2.1 It is accepted that the passing of component parts of a wiring system (such as cables, tray and trunking) through ceilings, floors, walls and the like is unavoidable. However, in order to reduce the possibility of spread of fire and the products of combustion to a minimum, it is important that any opening remaining around such parts where they penetrate an element of the building construction is made good to at least the same standard as that which existed prior to the penetration being made (see Figure 6.1).

The requirements to seal openings apply not only for elements of building construction that subdivide a building into fire compartments (in fact many dwellings consist of only a single fire compartment), they also apply where a wiring system penetrates an element of building construction that has a specified fire resistance and is relied on for structural stability. This is because the purpose of sealing is not only to preserve fire separation between areas of a building; it is also to preserve the structural stability of the premises, such as the loadbearing capacity of floors, in the event of fire.

Even if only a single fire resisting lining of an element (such as plasterboard) is penetrated, the opening should be sealed, as the lining might be responsible for providing most of the fire resistance of the element, with the interior (such as timber studs or webbed 'I' joists) having only a short survival time if exposed to fire.

527.2.1.1
527.2.1.2 Temporary sealing arrangements should be provided, as and where required, throughout the construction of an installation. Also, wherever any sealing of penetrations around wiring systems is disturbed whilst work is being carried out, it should be reinstated as soon as practicable.

527.2.4 Any sealing arrangement being used to meet the requirements of Regulation 527.2.1 or 527.2.1.1 as described above should:

(a) be capable of resisting external influences to the same degree as the wiring system component with which it is being used;
(b) be equally resistant to the products of combustion and to penetration by water as the element of building fabric which has been penetrated;
(c) be compatible with the material of the wiring system with which it is in contact such that neither is degraded;
(d) permit sufficient thermal movement (through expansion and contraction) of the wiring system such that the sealing remains effective; and
(e) be sufficiently robust such that it can withstand the stresses likely to occur as a result of damage to the supports of the wiring system caused by fire. This condition is considered to be met where cable cleats, clips or other forms of support are installed within 750 mm of the seal and are capable of withstanding the strain following any such collapse.

▼ Figure 6.1 External and internal sealing arrangements

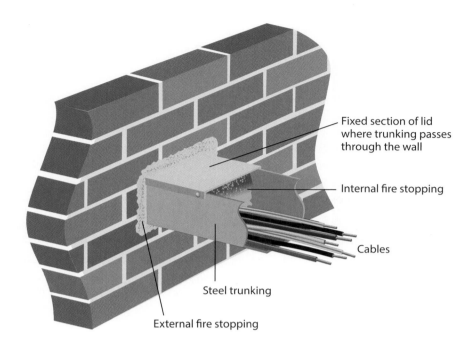

Fixed section of lid where trunking passes through the wall

Internal fire stopping

Cables

Steel trunking

External fire stopping

Furthermore, any such sealing arrangement and the wiring system itself, unless fully resistant to moisture, should be protected from dripping water which might travel along the wiring system to the seal or otherwise allow water to collect on or around the seal.

6.2.2 Internal sealing arrangements

527.2.2
527.2.3 With the exception of conduit, trunking or ducting systems classified as non-flame propagating in accordance with the relevant product standard, and having an internal cross-sectional area not exceeding 710 mm^2, any wiring system passing through an element of the building fabric having a specified fire resistance should be internally sealed to provide the same degree of fire resistance as the element of the building fabric being penetrated, in addition to the external sealing requirements described above (see Figure 6.1).

In practical terms this means that standard conduit of diameter 32 mm or less, or trunking or ducting not exceeding dimensions 25 × 25 mm or similar, will not require internal sealing.

6.2.3 Material for sealing wiring system penetrations

It is advisable to use products that have appropriate fire test or assessment evidence to cover the end-use configuration for sealing wiring system penetrations. Just because a product carries a 'fire rating' does not mean it is a suitable firestopping product. Details of products that have been comprehensively fire tested and third party certificated can be found in the Association for Specialist Fire Protection (ASFP) 'Red Book' – *Fire Stopping: Linear joint seals, penetration seals & small cavity barriers, 4th Edition*.

Examples of materials that can be used for external sealing include intumescent mastics, compounds, metal sleeves and fire-resistant sponge-filled multi-service boxes.

Internal sealing can often be achieved using intumescent material, which could be in the form of mastic that can be pumped in, or as pillows for use with trunking and cable tray.

Temporary sealing arrangements, where necessary, usually require the use of removable firestopping products, such as intumescent pillows.

Reinstatement of existing seals that have been disturbed should use the same types of materials/components as originally used, as mixing and matching of components is not supported by manufacturers' fire test data. If the original seal cannot be identified or sourced, the whole seal should be replaced.

6.2.4 Flush-mounted accessories penetrating stud partitions

Back boxes of flush-mounted accessories (for example, switches and socket-outlets) generally incorporate knockout sections significantly larger than the cables passing through them. This makes these openings very permeable in a fire after the accessory faceplate has been destroyed by the heat, allowing hot gases into the cavity of stud partition walls much more rapidly than the plasterboard.

Where flush-mounted accessories penetrate each face of a 30 minute fire separating or loadbearing plasterboard lined wall within the same cavity space (see Figure 6.2), the back box of each accessory should incorporate integral fire protection or be fitted with a proprietary fire protection pad, unless evidence of the fire resistance performance of the accessories is available (Figure 6.3 shows examples of where protection pads are fitted).

Such back boxes or protective pads must have evidence of performance to demonstrate that they have the ability to maintain the fire separation capability of a wall for 30 minutes, were they to be tested to BS 476: Part 21 or BS EN 1365: Parts 1 and 2 (loadbearing), or BS 476: Part 22 or BS EN 1364: Parts 1 and 2 (non-loadbearing), with plastic accessories fitted in both linings.

▼ **Figure 6.2** Examples of where accessories are and are not back-to-back in the same cavity space

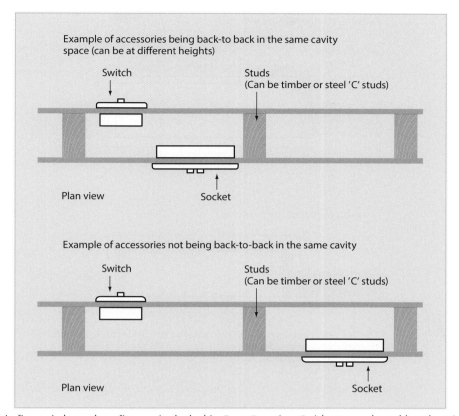

This figure is based on figures included in Best Practice Guide 5, produced by Electrical Safety First in association with leading industry bodies, including the IET.

▼ **Figure 6.3** Examples of fire protection pads fitted to accessory back boxes

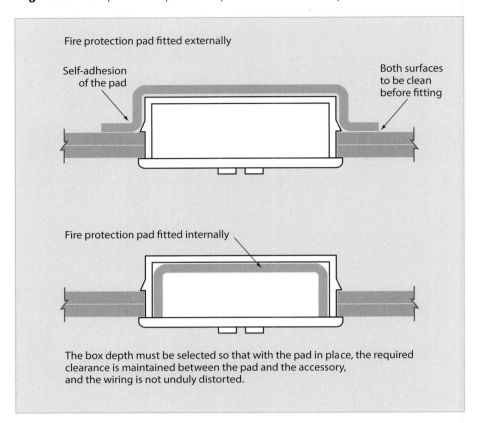

Fire protection pad fitted externally

Self-adhesion of the pad

Both surfaces to be clean before fitting

Fire protection pad fitted internally

The box depth must be selected so that with the pad in place, the required clearance is maintained between the pad and the accessory, and the wiring is not unduly distorted.

This figure is based on figures included in Best Practice Guide 5, produced by Electrical Safety First in association with leading industry bodies, including the IET.

Regulation 642.3

6.2.5 Inspection

It should be remembered that Regulation 642.3 (vii) calls for inspection, during the installation process and on completion of work, to verify the presence of fire barriers, seals and other measures provided for protection against thermal effects.

See also Appendices D and E of this Guidance Note for requirements relating to fire-stopping in the Building Regulations of England and Wales and of Scotland, respectively.

For installations in Northern Ireland reference should be made to the Northern Ireland Building Regulations website: www.buildingregulationsni.gov.uk.

Safety services   **7**

7.1   General

560.1 The requirements of Chapter 56 are applicable to supplies to services which need to remain functional during a fire. These would include services such as booster pumps for sprinkler systems and risers; carbon monoxide (CO) detection/alarm systems; emergency lighting; fire detection/alarm systems; fire evacuation systems; fire services communication systems; lifts designated for use by the fire service; industrial safety systems; life-critical medical systems; and smoke extraction equipment. The requirements are not applicable to installations in hazardous areas, for which reference should be made to BS EN 60079-14, or to electrical standby systems.

Note: Any wiring connecting a self-contained emergency luminaire to the main supply is not considered to be an emergency lighting circuit.

7.2   Classification and changeover times

560.4.1 Electrical safety service supplies may be:

(a) non-automatic – in which case the service is activated manually by an operator; or

(b) automatic – where operation does not require operator input.

Regulation 560.4.1 classifies automatic supplies by the maximum changeover time as follows:

(a) no-break – where a continuous supply is produced subject to specified conditions on, for example, voltage and frequency;

(b) very short break – supply is available within 0.15 s;

(c) short break – supply is available within 0.5 s;

(d) normal break – supply is available within 5 s;

(e) medium break – supply is available within 15 s; and

(f) long break – supply is available in a time exceeding 15 s.

560.4.2 It is a requirement that essential equipment for safety services is compatible with these system classification changeover times.

Safety services can be expected to be called upon at all material times, that is when the premises in question are likely to be occupied by persons and/or livestock and hence those occupants may be at risk as a result of fire and/or the failure of the normal supply arrangements.

560.5.2 Where a safety service is required to remain fully operational under fire conditions, the safety source of supply must be sufficiently robust to allow the safety service being supplied to perform its function correctly.

Taking emergency lighting by way of example, BS EN 1838:2013 specifies a minimum duration for emergency escape lighting of 1 hour (see Clauses 4.2.5 and 4.3.5) whilst BS 5266-1:2016 specifies a minimum duration of 3 hours in premises where persons are likely to sleep or may be unfamiliar with the layout of the premises, or may be infirm, such as hospitals, nursing homes, schools, colleges and hotels. Any specification for such emergency lighting would have to take these duration minima into account.

Furthermore, any installed equipment should, either by its construction or as a result of the way in which it has been installed, have a sufficient degree of fire resistance to enable it to remain operational for the prerequisite time.

560.5.3 It is preferred that any measures providing fault protection do not result in automatic disconnection under first fault conditions. For example, Clause 12.2.1 of BS 5839-1:2017 requires fault indication at the control and indicating equipment (which is typically sited in the entrance lobby of a building, security office or some such similar location likely to be occupied at all material times) to occur within 100 seconds if the following were to occur:

(a) a short-circuit or open circuit of any
  (i)   circuits supplying fire alarm devices;
  (ii) wiring between a power supply and any equipment which it supplies that is housed in a separate enclosure;
  (iii) wiring between the control and indicating equipment and any remote control/indicating equipment, such as repeater panels; and
  (iv) wiring between control equipment and any equipment in a separate enclosure for the transmission of signals to an alarm receiving centre
(b) any earth fault that would be capable of preventing the fire alarm system from operating in accordance with the recommendations of BS 5839-1.

If a safety supply is derived from an IT system, continuous insulation monitoring should be provided such that an audible and visual indication is given when a first fault occurs.

7.3   Sources for safety services

560.6.2 Any equipment forming part of a source for a safety service must be installed as fixed
560.6.3 equipment in a location that is only accessible to skilled or instructed persons.

7.3.1   Generators

560.6.4 Consideration should be given at the design stage to ensure that a source of supply such as an oil-powered generator is installed such that any gases, vapours, smoke or fumes emitted cannot access parts of a building which are intended to be routinely occupied by persons. This will require the provision of suitable ventilation arrangements including, in some cases, the installation of extraction plant.

560.6.5 The use of two supplies derived from a public supply network should not be selected as safety supply services unless it can be confirmed that they will not be prone to fail at the same time. As such, this would require discussion between the designer and the distributor to ensure that the supplies were appropriately configured.

560.6.6
560.6.7 Any safety source should be capable of supporting at least all of the related safety services. However, a safety source may supply additionally other electrical loads within **314.1** the premises where this does not detrimentally impact upon the supply required by the safety services. As such, faults occurring on a non-safety service supplied from a safety source on failure of the normal supply arrangements must not result in the disconnection of a safety service. In practice, care must be exercised when arranging the protective measures for all circuits to ensure that the protective device closest to a fault will operate, minimising disruption to and disconnection of the healthy circuits and preventing danger being caused by the loss of safety services. Consideration should also be given to arrangements for load-shedding to ensure that safety services are given priority in the event of failure of the normal supply.

560.6.8 Where a generator providing an alternative source for a safety supply is used, it is necessary to ensure that the relevant requirements of Section 551 are met. Such requirements include that:

551.4.3.2.1 **(a)** the generator must have its own earthing arrangements independent of those derived from the earthed point of a TN system forming part of a public supply, such that the generator may be operated safely in the event of loss of the public supply earthing arrangement

(b) precautions must be taken to prevent the generator being connected in parallel with the normal supply derived from a public supply network (unless specifically designed for this purpose). This may be achieved by employing one or more of the following measures:

551.6.1 **(i)** electrical, mechanical or electromechanical interlocks;
(ii) a system of locks with a single transferable key;
(iii) a three-position break-before-make changeover switch;
(iv) an automatic changeover switch with suitable interlock; or
(v) other suitable means which prevent parallel operation.

551.6.2 For TN-S system installations in situations where the neutral is not isolated, care must be taken to ensure that RCDs are so positioned that they do not operate inadvertently as a result of any parallel neutral-earth path. Consideration should also be given to isolation of the installation neutral from the public supply neutral in TN system installations as a means of avoiding disturbances arising from induced voltages caused by lightning being introduced into the installation.

560.6.9
434.5.2 Where a generator providing an alternative source for a safety supply is capable of operating in parallel with the normal supply, short-circuit and fault protection must be **411.3.2** effective irrespective of how the installation is supplied . For example, the designer **551.7.2** must confirm that the measures employed for protection against thermal effects (Regulation 434.5.2) and protection against electric shock under fault conditions (Regulation 411.3.2) will remain effective at all times under all modes of operation. If the generator is placed on the supply side of all the protective devices for the final circuits of the installation, no special measures are required.

If, however, the generator is placed on the load side of all the overcurrent protective devices for a final circuit of the installation, the conductors of the final circuit in question should meet the following requirement:

$$I_z \geq I_n + I_g$$

where:

I_z is the current-carrying capacity of the final circuit conductors

I_n is the rated current of the protective device of the final circuit

I_g is the rated output current of the generating set.

All of the other requirements of Regulation 551.7.2 must also be met.

560.6.13 Any generator forming a part of an electrical source for a safety service should comply with the relevant Clauses of BS 7698-12, *Reciprocating internal combustion engine driven alternating current generating sets. Emergency power supply to safety devices.*

7.4   Circuits

560.7.1 Circuits supplying safety services should be independent of other circuits such that a fault, modification work or maintenance on one type of circuit does not have a detrimental effect on any other. This is generally achieved by the use of independent wiring systems for the safety services, or suitably compartmentalised wiring systems.

560.7.2 Circuits should, wherever possible, be routed to avoid their passing through locations that present a fire risk. However, where this cannot be avoided the cables must be fire-resistant. Circuits of safety services should not be routed through areas of a building that are exposed to an explosion risk.

560.7.3
433.3.3 Generally, circuits for safety services should be provided with overload protection. However, Regulation 560.7.3 refers to Regulation 433.3.3 and permits the omission of overload protection where the loss of supply under such conditions could cause a greater hazard. Where protection against overload is omitted, the occurrence of an overload should be indicated in some manner such as warning lights and/or supervisory buzzers.

560.7.4
560.7.6 Given the importance of maintaining the supply to safety services, care should be taken at the design stage to ensure that an overcurrent in one circuit has no detrimental effect, such as unwanted disconnection of other circuits supplying safety services. Further, where equipment is supplied from two different circuits, they should be so arranged that the supply to the equipment is maintained if a fault were to occur on either of the circuits. A fault occurring on either circuit should not detrimentally affect the measures for protection against electric shock or the correct operation of the remaining circuit.

It should be noted that Clause 15.5 (a) (ii) of BS 5839-6:2013 permits domestic fire detection/alarm devices forming part of a grade D system (as defined in BS 5839-6) to be supplied from a 'separately electrically protected or regularly used local lighting circuit'. Reference should also be made within this Guidance Note to Section 1.19 of Appendix B (for installations in England and Wales) and Clauses 2.11.9 and 2.11.10 of Appendix E (for installations in Scotland).

560.7.7 It is necessary for cabling other than metallic screened fire-resistant cables forming part
528.1 of a safety circuit to be effectively and reliably separated from the cabling of other circuits in accordance with the general requirement given in Regulation 528.1 and any specific requirements of the relevant British Standard (such as, in the case of fire alarm systems, those given in Clause 26.2.k of BS 5839-1:2017).

Clause 26.1 of BS 5839-1:2017 states that "the circuits of fire alarm systems need to be segregated from the cables of other circuits to minimize any potential for other circuits to cause malfunction of the fire alarm system".

560.7.8 Regulation 560.7.8 prohibits the installation of any wiring other than that directly associated with the fire rescue service lift within a lift shaft or any other flue-like opening.

7.5 Information to be made available

Chapter 56 contains a number of requirements regarding information that must be made available and remain accessible to the occupants of a building and indeed emergency services who may be required to attend the premises in the case of fire. These are summarised here:

560.7.5 **(a)** Any switchgear/controlgear associated with the supply of a safety service should be clearly identified. Any such switchgear should also be located in a plant room or similar such that it is only accessible to skilled or instructed persons

560.7.9 **(b)** A general schematic diagram and details of the supply arrangements of the electrical safety services should be posted in close proximity to the relevant distribution board. The diagram need not be overly complex, but its content must be maintained as and when changes to the supply are made, so that it remains an accurate reflection of the system as installed

560.7.10 **(c)** Copies of all relevant installation drawings should be posted at the origin of the installation such that the location of the following items can be pinpointed:

560.7.10
   (i) all electrical control equipment including distribution boards;
   (ii) items of safety equipment with the relevant final circuit designation. The particulars and purpose of the equipment should be evident; and
   (iii) any special switching and monitoring equipment such as area switches and visual and/or acoustic warning devices relating to the safety power supply

560.7.11 **(d)** A list should be provided detailing all current-using equipment which is permanently connected to a safety power supply. Relevant information relating to power, rated voltage, current, starting current and duration should be provided

560.7.12 **(e)** The operating instructions as appropriate for the particular installation conditions should be provided for any items of safety equipment and electrical safety services.

7.6 Wiring systems

560.8.1 Wiring systems forming part of a safety service that is required to operate under fire conditions need to comply with Regulation 560.8.1.

To meet the requirements of that regulation, the cables should comply with either:

(a) BS EN 50362 or BS 8491 if over 20 mm overall diameter; or
(b) BS EN 50200 if the overall diameter is less than or equal to 20 mm, in terms of fire resistance.

Additionally, in either case they should also comply with BS EN 60332-1-2 for resistance to flame.

Other cables deemed to offer the necessary degree of fire and mechanical protection may also be used, in which case reference should be made to the recommendations of the appropriate British Standard.

In the case of fire detection and alarm systems in buildings the requirements of BS 5839-1 apply and in the case of emergency lighting installations the requirements of BS 5266-1 should be met.

BS 5839-1:2017 recommends (amongst other things) that the wiring system for certain parts of the fire alarm installation should meet specified fire resistance requirements exceeding those referred to in the relevant product standards. BS 5266-1:2016 gives similar recommendations for the wiring system of certain parts of emergency lighting systems having a central battery source.

For fire alarm systems, the recommendations apply to all parts of the critical signal paths, the extra low voltage supply from the external power supply units, and the final circuit providing the low voltage mains supply to the system. Depending on the situation, Clause 26.2 of BS 5839-1:2017 recommends the use what it calls 'standard' fire resisting cable or 'enhanced' fire resisting cables. In both cases Clause 26(b) states that the cables should comprise one of the following types:

(a) mineral insulated copper sheathed cables, with an overall polymeric covering, conforming to BS EN 60702-1, with terminations conforming to BS EN 60702-2;
(b) cables that conform to BS 7629-1;
(c) cables that conform to BS 7846; or
(d) cables rated at 300/500 V (or greater) that provide the same degree of safety to that afforded by compliance with BS 7629-1.

Standard fire resisting cables should meet the PH 30 classification when tested in accordance with BS EN 50200 and additionally the 30 min survival time when tested in accordance with Annex E of that standard. Clause 26.2(c) of BS 5839-1:2017 indicates that these cables should adequately resist the effects of fire in most fire detection and alarm systems. Enhanced fire resisting cables should meet the PH 120 classification when tested in accordance with BS EN 50200 and the 120 min survival time when tested in accordance with BS 8434-2.

For central battery emergency lighting systems, the recommendations for the wiring system fire resistance mentioned above apply to cables or cable systems used for the connection of an emergency escape lighting luminaire to the standby power supply. In this context Clause 8.2.2 BS 5266-1:2016 gives the following classifications for types of cables and cable systems:

(a) **Emergency lighting cables with an inherently high resistance to attack by fire.** These should have a duration of survival of 60 min when tested in accordance with BS EN 50200:2015 and 30 min when tested in accordance with BS EN 50200:2015, Annex E. They should conform to BS EN 60702-1, with terminations conforming to BS EN 60702-2, to BS 7629-1 or to BS 7846.
(b) **Enhanced emergency lighting cables with an inherently high resistance to attack by fire.** These should have a duration of survival of 120 min when tested in accordance with BS EN 50200:2015 and 120 min when tested in accordance with BS 8434-2. They should conform to BS EN 60702-1, with terminations conforming to BS EN 60702-2, to BS 7629-1 or to BS 7846.

(c) **Emergency lighting cable systems with an inherently high resistance to attack by fire.** These should comprise fire-resistant single core or multi-core cables enclosed in screwed steel conduit, such that the cable system has a duration of survival of 60 min. The cable should meet the requirements of IEC 60331-3 for a flame application time of 60 min.

(d) **Enhanced emergency lighting cable systems with an inherently high resistance to attack by fire.** These should comprise fire-resistant single core or multi-core cables enclosed in screwed steel conduit, such that the cable system has a duration of survival of 120 min. The cable system should meet the requirements of IEC 60331-3 for a flame application time of 120 min.

Generally speaking, for most fire alarm installations, the use of 'standard fire resisting cables' is considered to provide sufficient resistance to the effects of fire by BS 5839-1:2017. Similarly, for most central battery emergency lighting systems, the use of cables and cable systems with an 'inherently high resistance to attack by fire' is considered by to provide sufficient resistance to the effects of fire BS 5266-1:2016.

Again, generally speaking, the use of 'enhanced fire resisting cables' (for fire alarms) and 'enhanced cables/cables systems with 'inherently high resistance to attack by fire' (for central battery emergency lighting systems) is recommended by BS 5839-1:2017 and BS 5266-1:2016, respectively, in cases where the designer, specifier or regulatory authority, on the basis of a fire risk assessment that takes fire engineering considerations into account, deems this to be necessary. Such applications might include, for example:

(a) unsprinklered buildings (or parts of buildings) in which the fire strategy involves evacuation of the occupants in four or more phases;

(b) unsprinklered buildings of greater than 30 m height; and

(c) unsprinklered premises and sites in which a fire in one area could affect cables associated with areas remote from the fire, in which it is envisaged that people will remain in occupation during the course of the fire.

The relevant content of BS 5839-1 and BS 5266-1 should be consulted for further details.

In all cases the methods of jointing and support for the fire alarm and emergency lighting cables should be as described in the relevant Clauses of BS 5266-1 relating to minimization of the probability of early failure in the event of fire and the labelling of junction boxes to avoid confusion with other services.

Clause 26.2.f of BS 5839-1:2017 specifically precludes the use of plastic cable clips, ties and trunking as the sole means of supporting cables that forms part of the wiring of a fire alarm system both as a means of maintaining circuit integrity and to minimise the hazard presented by unsecured cables under fire conditions. The preclusion on the use of plastic clips, ties and the like as the sole means of cable support also appears in Clause 8.2.3 of BS 5266-1:2016.

It should be noted that minimum conductor sizes are stated for cables in both emergency lighting and fire alarm systems, including those used for supplies as an aid to maintaining circuit integrity as follows:

(a) 1 mm^2 for fire alarm systems (Clause 26.2j of BS 5839-1:2017 refers)

(b) 1.5 mm^2 for emergency lighting (Clause 8.2.7 of BS 5266-1:2016 refers).

Sect 528 Attention should be paid to the need to separate/segregate any cables or bus systems which might adversely affect the operation of the safety service, in accordance with the general requirements of Section 528 (of BS 7671) and any particular requirements of the appropriate British Standard. For fire alarm systems refer to Clause 26.2n of BS 5839-1:2017 and for emergency lighting installations see Clause 8.2.6 of BS 5266-1:2016.

560.9
560.10 Care should be taken to ensure that any wiring that forms part of the supply to the safety service is installed in such a way as to minimise the risk of short-circuit or earth fault, or the risk of fire or danger to persons. In some cases and with some types of cable, this may require some thought to be given to the routes taken by cables or the installation of additional mechanical protection. Again, reference should be made to the appropriate British Standard, and any particular recommendations therein should be taken into consideration.

560.9
560.10
560.11 For life safety and firefighting applications, Regulation 560.11 requires that power and control cable systems that are required to maintain their circuit integrity shall comply with BS 8519, *Selection and installation of fire-resistant power and control cable systems for life safety and fire-fighting applications – Code of practice*.

As well as giving guidance and recommendations on the selection and installation of cable systems, BS 8519 gives specific recommendations for electrical system design, and also gives recommended limits for survival times, for installations falling within the above description.

BS 8519 does not give recommendations for those installations covered in the following standards but makes reference to these standards in an informative capacity:

(a) BS 5839, *Fire detection and fire alarm systems for buildings*.
    (i) Part 1, *Code of practice for design, installation, commissioning and maintenance of systems in non-domestic premises*
    (ii) Part 8, *Code of practice for the design, installation, commissioning and maintenance of voice alarm systems*
    (iii) Part 9, *Code of practice for the design, installation, commissioning and maintenance of emergency voice communication systems*, and

(b) BS 5266-1, *Emergency lighting. Code of practice for the emergency escape lighting of premises*.

Special installations and locations 8

8.1 Introduction

Part 7 Although the general requirements of BS 7671, including those relating to protection against fire, will apply to all electrical installations, these may be added to or amended by more specific requirements considered to be appropriate in the case of special installations or locations.

BS 7671 now contains requirements for twenty types of special installation or location. Some of these sections contain specific additional requirements for protection against fire as follows:

Sect 705 **(a)** Agricultural and horticultural premises
Sect 710 **(b)** Medical locations
Sect 711 **(c)** Exhibitions, shows and stands
Sect 712 **(d)** Solar photovoltaic (PV) power supply systems
Sect 715 **(e)** Extra-low voltage lighting installations
Sect 740 **(f)** Temporary electrical installations for structures, amusement devices and booths at fairgrounds, amusement parks and circuses
Sect 753 **(g)** Heating cables and embedded heating systems.

These are considered below.

8.2 Agricultural and horticultural premises

8.2.1 The risks

Sect 705 Agricultural premises frequently contain buildings that may be used for a wide range of activities which could present or contribute to a risk of fire.

Given the characteristic high level roofs/ceilings found in many storage buildings on farms, the use of projecting beam flood and spot lights is commonplace. Parts of the premises may be used for storage, refuelling, maintenance and repair of vehicles and machinery. Whilst storage of most fuels in itself is not considered too hazardous, handling or refuelling may well be. The types of maintenance and repair activities being carried out could reasonably include hot work such as welding or brazing. Other buildings may be used for the large-scale storage of fertilisers, (which can, under fire conditions, act as oxidising agents); flammable feeds; and bedding materials such as hay and straw, which in themselves can present a fire risk if incorrectly stored even without the presence of an electrical installation.

Agricultural and horticultural premises, in particular areas used for storage of edible items, are likely to attract rodents such as mice and rats which are known to gnaw on cables, causing damage to the insulation, thereby increasing the risk of short-circuit or earth fault. These in turn could produce arcing sufficient to present a risk of ignition.

However, to make matters worse it can be quite common for buildings in agricultural premises to be used for any number of the purposes mentioned above at the same time, presenting a combination of flammable or combustible materials, oxidising agents, raised temperatures and the presence of unshielded arcs and sparks.

131.1 It should be remembered that any livestock present on a farm must also be safeguarded from excessive temperatures likely to be caused by heating appliances or lighting.

8.2.2 Radiant heaters

705.422.6 Any electrical heating appliances which are employed in areas used for housing livestock should meet the requirements of BS EN 60335-2-71 *Household and similar electrical appliances. Safety. Particular requirements for electrical heating appliances for breeding and rearing animals*. Any heaters installed in such locations should be installed in accordance with the specified clearance from the livestock or combustible materials stated by the manufacturer. In the absence of any such recommendation a minimum clearance of 0.5 m from livestock or combustible materials must be maintained.

8.2.3 Employment of RCDs as a measure for protection against fire

705.422.7 For additional fire protection in some circumstances, the use of RCDs having a rated residual operating current not exceeding 300 mA which disconnect all live conductors as a measure for protection against fire. Regulation 705.422.7 also accepts the use of 'S' type or time delay RCDs for fire protection purposes where the circuit being so protected does not supply socket-outlets.

It is highly likely that most circuits within an agricultural or horticultural installation will already be subject to RCD protection as a means of providing the protective measure of automatic disconnection of supply. Where this is the case, RCDs providing automatic disconnection of supply can also be deemed to provide the required protection against fire.

8.2.4 Additional requirements for ELV circuits

705.422.8 In any location where a risk of fire is considered to exist, any conductors of a circuit supplied at extra-low voltage should be provided with mechanical protection by means of barriers or enclosures such that a level of protection against ingress of IPXXD or IP4X is achieved, or be otherwise installed in a cable management system (i.e. conduit, trunking or similar) made from insulating materials.

It should be noted that type H07RN-F (BS EN 50525-2-21) heavy duty rubber industrial flexible cables meet the requirements of this regulation without the need for any additional measures being applied.

8.2.5 Rodent activity

705.513.2
705.522.10 Whilst Regulation 705.513.2 requires electrical equipment to be placed beyond the reach of any livestock that are likely to be present, this requirement relates only to animals which are intended to be present and not to vermin such as rats, mice or

squirrels. As such, Regulation 705.522.10 contains a particular requirement that special attention be given to the presence of rodents.

Besides the risk of cable damage through gnawing as mentioned earlier, rodents are known to nest on top of cable runs and within inadequately sealed enclosures, which again in itself may present a risk of fire.

Particular attention should be paid to the sealing arrangements applied to ducts and other cableways to prevent access for animals. Any cables considered to be so placed as to be vulnerable to the gnawing activities of rodents should be suitably protected from such actions by the use of steel trunking, conduit and accessory boxes, or possibly high impact plastic conduit, trunking etc. It is unlikely that further protection would be required for steel wire armoured or mineral insulated cables, although, given sufficient time, even these types of cable would not be beyond suffering serious damage.

8.2.6 Luminaires

Sects 559,714 and 715
Chap 42
Luminaires should be installed in accordance with the general requirements contained in Sections 559, 714 and 715, and Chapter 42 as appropriate (see Chapter 3 of this Guidance Note).

When luminaires are being selected for use in agricultural or horticultural premises consideration must be given to:

(a) the degree of protection offered against ingress of dust, solid objects and moisture, an IP rating of IP54 being a reasonable minimum standard

(b) their suitability for installation directly attached to a normally flammable surface, such luminaires may be marked $\boxed{\nabla\!\!\!F}$. However with the publication of BS EN 60598-1 2008, luminaires suitable for direct mounting on normally flammable surfaces have no special marking and only luminaires not suitable for mounting on normally flammable surfaces are marked with a symbol $\boxed{\text{🔥🔥}}$ $\boxed{\text{🔥}}$. Reference should be made to Section 3.2.4 of this Guidance Note for further guidance on symbols used in luminaires.

8.3   Medical locations

710.512.2.1
The section makes an important requirement for protection against fire in medical locations, in that electrical outlets and accessories should be installed at least 200 mm from any medical gas outlet or flammable gas outlet to minimise the risk of ignition. This applies where electrical outlets and accessories are to be installed below gas outlets.

8.4   Solar photovoltaic (PV) power supply systems

Sect 712
Unlike most electrical products, PV modules and wiring do not have an overall enclosure to contain arcs and fires resulting from faults, and many PV systems operate at DC voltages very capable of sustaining DC arcs.

Furthermore, because the DC side of a PV system is a current-limiting generating set, the protective measure Automatic Disconnection of the Supply is almost never used for protection against electric shock on that side. Instead, virtually all systems use the protective measure of double or reinforced insulation.

712.522.8.1 For these reasons, Regulation 712.522.8.1 requires that PV string cables, PV array cables and PV DC main cables shall be selected and erected in such a way that the risk of earth faults and short-circuits is kept to a minimum. For example:

(a) single-core (rather than multicore) non-metallic sheathed cables should be used; and

(b) it is advisable not to conceal the cables in walls or otherwise hide them in the building structure, as this would make it difficult to detect mechanical damage to the cables.

Sect 526 Faulty electrical connections are another potential cause of arcing on the DC side. It is **526.1** therefore essential that the correct means of connection, complying with Section 526 of BS 7671, are used and that all means of isolation and switching are appropriately rated and suitable for DC service.

In a PV array formed from a number of strings, particular attention is needed to ensure protective measures against overloaded string cables and excessive module reverse currents, both of which can present a considerable fire risk.

IET CoP More detailed guidance will be found in the *IET Code of Practice for Grid Connected Solar Photovoltaic Systems*.

8.5 Extra-low voltage lighting installations

Sect 715 Section 715 applies to extra-low voltage lighting installations and Regulation 414.4.5 **414.4.5** makes it clear that basic protection such as insulation is not required in normal dry conditions for:

(a) SELV circuits having nominal voltage not exceeding 25 V AC or 60 V DC; and

(b) PELV circuits having nominal voltage not exceeding 25 V AC or 60 V DC where the exposed-conductive-parts and/or live parts are connected by a protective conductor to the main earthing terminal.

Sect 715 Where both live conductors of an extra-low voltage lighting system are uninsulated, **414.4.5** the lighting system should comply with BS EN 60598-2-23, or the conductors should **715.422.107.1** be supplied from a transformer or convertor the power of which does not exceed 200 **715.422.107.2** VA, or the conductors should be protected by a device meeting all the of the following requirements (given in Regulation 715.422.107.2):

(a) continuously monitors the power demand of the connected luminaires;

(b) capable of automatically disconnecting the supply circuit within 0.3 s in the event of a short-circuit or other failure which results in a power increase of more than 60 W;

(c) provides automatic disconnection when the circuit is operating with reduced power such as may occur as a result of lamp failure, gating control or similar, or any failure causing a power increase of more than 60 W;

(d) provides automatic disconnection on connection of the supply circuit in the event of any failure which results in a power increase of more than 60 W; and

(e) the device should be fail-safe.

8.6   Exhibitions, shows and stands

711.422.4.2   BS 7671 considers the particular issues present in electrical installations for exhibitions, shows and stands. In the case of protection against fire, Regulation 711.422.4.2 contains the following requirements:

(a) any lighting equipment or other equipment being used within the location having a high surface temperature under conditions of normal operation should be so placed or guarded that it does not constitute a fire hazard;

(b) any signs or similar should be sufficiently robust in terms of heat resistance, mechanical strength, electrical insulation employed and ventilation such that they can adequately withstand heat likely under normal operating conditions; and

(c) where a stand contains a number of items of electrical equipment, lamp types likely to result in the production of excessive levels of heat should not be installed unless ventilation sufficient to avoid overheating problems is provided.

711.559.5   There is also a requirement for luminaires installed within arm's reach (2.5 m) of floor level to be securely fixed and suitably mounted and/or guarded to prevent a risk of injury, including burns or ignition of materials from occurring.

8.7   Temporary electrical installations for structures, amusement devices and booths at fairgrounds, amusement parks and circuses

740.422.3.7   Section 740 contains a specific requirement that any motor which is either automatically or remotely controlled but which is not under continuous supervision should be fitted with an excess temperature cut-out that requires manual resetting.

This measure is intended to ensure that an examination of the motor and its immediate surroundings is made prior to the supply being reinstated after the cut-out has operated.

740.55.1.5   Regulation 740.55.1.5 requires that any luminaires and floodlights be so fixed and protected that they do not constitute an ignition hazard to materials in the location as a result of the effects of focusing or concentration of the heat that they emit in normal use.

8.8   Heating cables and embedded heating systems

Sect 753   Section 753 contains a specific requirement limiting the surface temperature of floors likely to be trodden on to a reasonable level of, for example, not more than 35 °C.

For additional information reference can be made to *CENELEC Guide 29 Temperatures Of Hot Surfaces Likely To Be Touched* which provides guidance for assessing the risk of a burn from unintentional contact with readily accessible surfaces of electrical equipment operating with a voltage of between 50 V and 1000 V AC and 75V and 1500V DC. The document establishes surface temperature limits, where such limits are required, and describes the maximum contact periods with a hot surface that a person may be subjected to without being exposed to a risk of burn.

753.424.101
753.424.102   For wall heating systems the heating units should be provided with a metal sheath, metal enclosure or fine mesh metallic grid connected to the protective conductor of the supply circuit. To meet the requirements of Chapter 42 (Thermal effects), care should be taken to prevent the heating elements creating high temperatures to parts

of the building fabric. This can be achieved by using heating units with temperature self-limiting functions or by separation with heat-resistant materials. This separation can be achieved by placing the heating element on a metal sheet, in metal conduit or at a distance of at least 10 mm in air from the ignitable structure.

753.522.1.3 Consideration should be given to the likely increase in ambient temperature where circuit wiring and control leads are installed in close proximity to heated surfaces.

GN7 Further coverage of electrical installations in special installations or locations can be found in Guidance Note 7.

Appendix

Safety service and product standards of relevance to this Guidance Note

BS or EN number	Title	References in this Guidance Note
BS 476-4:1970	Fire tests on building materials and structures. Non-combustibility test for materials	3.9
BS 476-12:1991	Fire tests on building materials and structures. Method of test for ignitability of products by direct flame impingement	3.2.13.9
BS 476-21:1987	Fire tests on building materials and structures. Methods for determination of the fire resistance of loadbearing elements of construction	6.2.4
BS 476-22:1987	Fire tests on building materials and structures. Method for determination of the fire resistance of non-loadbearing elements of construction	6.2.4
		3.9
BS 4884 series	Technical manuals	Introduction
BS 4940 series	Technical information on construction products and services	Introduction
BS 5266 series	Emergency lighting	4.2
BS 5266-1:2016	Emergency lighting. Code of practice for the emergency lighting of premises	4.2
		7.2
		7.6
		Appx E 2.10.5
BS 5266-7:1999	Lighting applications. Emergency lighting (also known as BS EN 1838:1999) (this standard has been superseded by BS 1838: 2013, listed above, but is still referenced in the Scottish Technical Handbooks referred to in this Guidance Note)	Appx E 2.10.5
BS 5446-2:2003	Fire detection and fire alarm devices for dwellings. Specification for heat alarms	Appx B Vol. 1, 1.4
		Appx B Vol. 2, 1.5
BS 5839-1:2017	Fire detection and fire alarm systems for buildings. Code of practice for system design, installation, commissioning and maintenance (this standard has been superseded by BS 5839-13:2013, listed below, but is still referenced in the Approved Documents and Scottish Technical Handbooks referred to in this Guidance Note)	Appx B
		Appx E 2.11.1 Non-dom

BS or EN number	Title	References in this Guidance Note
BS 5839-1:2017	Fire detection and fire alarm systems for buildings. Code of practice for system design, installation, commissioning and maintenance in non-domestic premises	7.2
		7.4
		7.6
BS 5839-3:1988	Fire detection and alarm systems for buildings. Specification for automatic release mechanisms for certain fire protection equipment	Appx E 2.1.14 Non-dom
		Appx E 2.2.9 Dom
BS 5839-6:2004	Fire detection and fire alarm systems for buildings. Code of practice for the design, installation and maintenance of fire detection and fire alarm systems in dwellings (this standard has been superseded by BS 5839-6:2013, listed below, but is still referenced in the Approved Documents and Scottish Technical Handbooks referred to in this Guidance Note)	Appx B
		Appx E 2.11.3 Dom
		Appx E 2.11.4 Dom
BS 5839-6:2013	Fire detection and fire alarm systems for buildings. Code of practice for the design, installation and maintenance of fire detection and fire alarm systems in domestic premises	7.4
BS 5839-8:2008	Fire detection and fire alarm systems for buildings. Code of practice for the design, installation, commissioning and maintenance of voice alarm systems (this standard has been superseded by BS 5839-8:2013, listed below, but is still referenced in the Approved Documents and Scottish Technical Handbooks referred to in this Guidance Note)	Appx B Vol. 2, 1.32
BS 5839-8:2013	Fire detection and fire alarm systems for buildings. Code of practice for the design, installation, commissioning and maintenance of voice alarm systems	Appx B Vol. 2, 1.32
BS PD 6504:1983	Medical information on human reaction to skin contact with hot surfaces	Chap 5
BS 6724:2016	Electric cables. Thermosetting insulated, armoured cables for voltages of 600/1000 V and 1900/3300 V, having low emission of smoke and corrosive gases when affected by fire	Chap 2
BS 7211:2012	Electric cables. Thermosetting insulated, non-armoured cables for voltages up to and including 450/750 V, for electric power, lighting and internal wiring, and having low emission of smoke and corrosive gases when affected by fire	Chap 2
BS 7273-1:2006	Code of practice for the operation of fire protection measures. Electrical actuation of gaseous total flooding extinguishing systems	Appx B Vol. 2, 1.38
BS 7273-2:1992	Code of practice for the operation of fire protection measures. Mechanical actuation of gaseous total flooding and local application extinguishing systems	Appx B Vol. 2, 1.38
BS 7273-4:2007	Code of practice for the operation of fire protection measures. Actuation of release mechanisms for doors	Appx B Vol. 2, 1.38

BS or EN number	Title	References in this Guidance Note
BS 7273-5:2008	Code of practice for the operation of fire protection measures. Electrical actuation of watermist systems (except pre-action systems)	Appx B Vol. 2, 1.38
BS 7540 series	Electric cables. Guide to use for cables with a rated voltage not exceeding 450/750 V	Chap 2
BS 7629-1:2015	Electric cables. Specification for 300/500 V fire resistant screened cables having low emission of smoke and corrosive gases when affected by fire. Multicore and multipair cables	Chap 2
		7.6
BS 7671:2018	Requirements for Electrical Installations. IET Wiring Regulations. Eighteenth Edition (for convenience this standard is referred to as BS 7671 in this Guidance Note)	Numerous
BS 7698-12:1998	Reciprocating internal combustion engine driven alternating current generating sets. Emergency power supply to safety devices (also known as ISO 8528-12:1997)	7.3
BS 7846:2015	Electric cables. Thermosetting insulated, armoured, fire-resistant cables of rated voltage 600/1000 V, having low emission of smoke and corrosive gases when affected by fire. Specification	Chap 2
		7.6
BS 8434-2: 2003+A2:2009	Methods of test for assessment of the fire integrity of electric cables. Test for unprotected small cables for use in emergency circuits. BS EN 50200 with a 930° flame and with water spray	7.6
BS 8491:2008	Method for assessment of fire integrity of large diameter power cables for use as components for smoke and heat control systems and certain other active fire safety systems	4.6
		7.6
BS EN 54-2:1997+A1:2006	Fire detection and fire alarm systems. Control and indicating equipment	Appx B Vol. 1, 1.6
		Appx B Vol. 1, 1.7
BS EN 54-4:1998	Fire detection and fire alarm systems. Power supply equipment	Appx B Vol. 1, 1.6
		Appx B Vol. 1, 1.7
BS EN 54-11:2001	Fire detection and fire alarm systems. Manual call points	Appx B Vol. 2, 1.31
		Appx E 2.11.1 Non-dom
BS EN 1364-1:2015	Fire resistance tests for non-loadbearing elements. Walls	6.2.4
BS EN 1364-2:2018	Fire resistance tests for non-loadbearing elements. Ceilings	6.2.4
BS EN 1838:2013	Lighting applications. Emergency lighting	7.2
BS EN 14604:2005	Smoke alarm devices	Appx B Vol. 1, 1.4
		Appx B Vol. 2, 1.6
BS EN 50085 series	Cable trunking systems and cable ducting systems for electrical installations	4.2
		4.3.4
		6.1

BS or EN number	Title	References in this Guidance Note
BS EN 50085-1:2005+A1:2013	Cable trunking systems and cable ducting systems for electrical installations. General requirements	4.4.3
BS EN 50200:2015	Method of test for resistance to fire of unprotected small cables for use in emergency circuits	4.6
		7.6
BS EN 50362:2003	Method of test for resistance to fire of larger unprotected power and control cables for use in emergency circuits	7.6
BS EN 60079-14:2014	Explosive atmospheres. Electrical installations design, selection and erection	4.3
		7.1
BS EN 60332-1-2:2004+A11:2016	Tests on electric and optical fibre cables under fire conditions. Test for vertical flame propagation for a single insulated wire or cable. Procedure for 1 kW pre-mixed flame	4.3.4
		4.4.3
		6.1
		7.6
BS EN 60332-3 series	Tests on electric and optical fibre cables under fire conditions. Test for vertical flame spread of vertically-mounted bunched wires or cables.	4.2
		4.3.4
		4.5
BS EN 60335-2-71:2003	Household and similar electrical appliances. Safety. Particular requirements for electrical heating appliances for breeding and rearing animals	8.2.2
BS EN 61439-6: 2012	Low voltage switchgear and controlgear assemblies. Particular requirements for busbar trunking systems (busways).	4.3.4
		6.1
BS EN 60529:1992+A2:2013	Specification for degrees of protection provided by enclosures (IP code)	4.3.3
BS EN 60598-1:2015	Luminaires. General requirements and tests	3.2.4
BS EN 60598-2-23:1997	Luminaires. Particular requirements. Extra low voltage lighting systems for filament lamps	3.2.3
		8.6
BS EN 60598-2-24:2013	Luminaires. Particular requirements. Luminaires with limited surface temperatures	4.3.1
BS EN 60670-1: 2005+A1:2013	Boxes and enclosures for electrical accessories for household and similar fixed electrical installations. General requirements	4.5
BS EN 60702 series		4.6
BS EN 60702-1:2002+A1:2015	Mineral insulated cables and their terminations with a rated voltage not exceeding 750 V. Cables	Chap 2
		4.3.4
		7.6

BS or EN number	Title	References in this Guidance Note
BS EN 60702-2:2002+A1:2015	Mineral insulated cables and their terminations with a rated voltage not exceeding 750 V. Terminations	7.6
BS EN 61034-2:2005+A1:2013	Measurement of smoke density of cables burning under defined conditions. Test procedure and requirements	4.2
BS EN 61184:2017	Bayonet lampholders	3.4
BS EN 61386 series	Conduit systems for cable management	6.1
BS EN 61386-1:2008	Conduit systems for cable management. General requirements	4.2
		4.3.4
		4.4.3
BS EN 61439-3:2012	Low-voltage switchgear and controlgear assemblies. Distribution boards intended to be operated by ordinary persons (DBO)	3.9.1
BS EN 61534 series	Powertrack systems	4.2
		4.3.4
		6.1
BS EN 61537:2007	Cable management. Cable tray systems and cable ladder systems	4.3.4
		6.1

A

68 Guidance Note 4: Protection Against Fire
© The Institution of Engineering and Technology

Appendix B

Approved Document B1:
Means of warning and escape

For Scotland see Appendix E of this Guidance Note.

For Northern Ireland see the Technical Booklets listed in Explanatory Note 8 to the Building Regulations (Northern Ireland) 2012 as amended. The publications include Technical Booklet E:2012 and other fire safety documents. The legislation and Technical Booklet can be found on the Department of Finance and Personnel's building regulations website: www.buildingregulationsni.gov.uk

For England and Wales, Approved Document B (Fire safety) is subdivided into two volumes. Volume 1 contains guidance applicable to dwellinghouses while Volume 2 contains guidance applicable to buildings other than dwellinghouses. Consequently, the guidance in this Appendix is similarly divided along these lines.

In the text below, Approved Document B is abbreviated to ADB.

At the date of printing of this Guidance Note, different versions of ADB apply in England and Wales, as follows:

(a) England:
- **(i)** Volume 1 – 2006 edition incorporating 2010, 2013 and 2017 amendments
- **(ii)** Volume 2 – 2006 edition incorporating 2007, 2010, 2013 and 2017 amendments

(b) Wales:
- **(i)** Volume 1 – 2006 edition incorporating 2010 and 2017 amendments
- **(ii)** Volume 2 – 2006 edition incorporating 2010, 2013 and 2017 amendments

Unless otherwise indicated the text in this appendix relates to both England and Wales.

Where dated versions of British Standards are referred to in the text below, these versions are the ones referred to in ADB, which in some cases are not the latest version of the British Standard concerned. Where this is the case, compliance with the relevant provisions of the Building Regulations will generally be achieved by following the corresponding provisions of the latest version of the British Standard concerned.

1.1   (ADB Vols. 1 and 2)

Section 1 of both volumes prescribes measures to be taken in buildings with respect to fire detection and alarm systems in order to give early warning in the event of fire.

Volume 1: Dwellinghouses

Note: Volume 1 is relevant to houses and the individual residential units of sheltered housing. For guidance applicable to the common parts of sheltered housing developments, flats, student accommodation and similar buildings, reference should be made to Volume 2.

1.2 (ADB Vol. 1)

It is recognised that the installation of either smoke alarms or automatic fire detection and alarm systems can significantly increase the level of safety by their raising the alarm automatically in the event of a fire breaking out. In most cases the guidance given in Volume 1 will be appropriate; however, a higher standard of protection might be necessary where occupants are at particular risk. As such, the approved document gives minimum recommendations for typical domestic situations.

1.3 (ADB Vol. 1)

All new dwelling houses should be provided with a fire detection and fire alarm system. This system should meet, as a minimum, the recommendations of BS 5839-6:2013 for a Grade D, LD3 standard.

BS 5839-6:2013 *Code of practice for the design, installation and maintenance of fire detection and fire alarm systems in dwellings* defines a Grade D system as 'A system of one or more mains-powered smoke alarms, each with an integral standby supply (the system may, in addition, incorporate one or more mains-powered heat alarms, each with an integral standby supply)' and Category LD3 as 'A system incorporating detectors in all circulation spaces that form part of the escape routes from the dwelling'.

1.4 (ADB Vol. 1)

Any smoke and heat alarms used should be mains-operated and have a standby supply from a battery or capacitor.

Smoke alarms should conform to BS EN 14604:2005 *Smoke alarm devices.*

Heat alarms should conform to BS 5446-2:2003 *Fire detection and fire alarm devices – Specification for heat alarms.*

1.5 (ADB Vol. 1)

A dwellinghouse is considered to be large if it has more than one storey and any such storey exceeds 200 m^2.

1.6 (ADB Vol. 1)

A large dwellinghouse having two storeys (not including any basement storey) should be fitted with a fire detection and fire alarm system which meets, as a minimum, the recommendations of BS 5839-6:2013 for a Grade B, Category LD3 system.

BS 5839-6:2013 defines a Grade B system as 'A fire detection and fire alarm system comprising fire detectors (other than smoke alarms and heat alarms), fire alarm sounders, and control and indicating equipment that either conforms to BS EN 54-2 (and power supply complying with BS EN 54-4) or to Annex C of this part (6) of BS 5839.'

A Category LD3 system is defined in BS 5839-6:2004 as 'A system incorporating detectors in all circulation spaces that form part of the escape routes from the dwelling.'

1.7   (ADB Vol. 1)

A large dwellinghouse having three or more storeys (not including any basement storey) should be fitted with a fire detection and fire alarm system which meets, as a minimum, the recommendations of BS 5839-6:2013 for a Grade A, Category LD2 system.

BS 5839-6:2013 defines a Grade A system as 'A fire detection and fire alarm system, which incorporates control and indicating equipment conforming to BS EN 54-2, and power supply equipment conforming to BS EN 54-4...' In the most part, such a system should be designed and installed in accordance with the relevant parts of BS 5839-1:2017. However, those aspects of the design/installation relating to audible alarm devices and audibility; fire alarm warnings for persons who are deaf or hard of hearing; manual call points; capacity of standby batteries; and radio linked systems, should be in accordance with the relevant parts of BS 5839-6:2013. For more precise information on this matter, refer to Clause 7.1 of BS 5839-6:2013.

A Category LD2 system is defined in BS 5839-6:2013 as 'A system incorporating detectors in all circulation spaces that form part of the escape routes from the dwelling, and in all rooms or areas that present a high fire risk to occupants ...'

1.8   (ADB Vol. 1)

Where alterations or additions are made to a domestic dwelling such that new habitable rooms are provided above ground floor level, or at ground floor level and there is no final exit from these newly provided rooms, it will be necessary to install a fire detection and alarm system. Smoke alarms will be required in the circulation spaces in accordance with Sections 1.10 to 1.18, guidance on which is given below.

The intention is that occupants of these new rooms will receive warning of a fire occurring in areas forming part of their escape route from the premises.

1.9   (ADB Vol. 1)

The fire detection and alarm systems in the individual dwellings in a sheltered housing scheme under the control of a warden or supervisor should be connected to a central monitoring point or alarm receiving centre. It should be possible for the person watching over the monitoring point for the sheltered housing complex to identify in which individual dwelling(s) a fire alarm has been raised.

It should be remembered that communal areas of a sheltered housing scheme should meet the recommendations of Volume 2 of Approved Document B, as should the sheltered accommodation of institutional/residential premises such as nursing accommodation, police section houses and the like.

1.10   (ADB Vol. 1)

In general the design and installation of the fire detection and alarm systems in dwellinghouses should be in accordance with the recommendations given in BS 5839-6:2013. However, the following points are emphasised in ADB Volume 1.

1.11   (ADB Vol. 1)

In order that an alarm is raised in the early stages of a fire, smoke alarms should be installed in circulation spaces such as corridors, hallways or landings between bedrooms, and those places where fires are most likely to start (that is, kitchens and living rooms).

1.12   (ADB Vol. 1)

At least one smoke detector should be installed on every storey.

1.13   (ADB Vol. 1)

An open plan situation where the kitchen area is not separated from stairways or circulation spaces by a door requires a compatible heat detector or heat alarm (in terms of the system to which it is to be connected) to be placed in the kitchen area, which should be interconnected to the smoke detector(s) installed in the circulation areas.

1.14   (ADB Vol. 1)

Where the alarm system consists of more than one detector/sounder unit, they should be linked so that detection of smoke or heat by one unit raises the alarm in all of them simultaneously. Any manufacturers' recommendations regarding the maximum number of units that can be linked should be observed.

1.15   (ADB Vol. 1)

The following should be observed when siting detectors on typical flat ceilings:

(a) in a circulation area there should be a smoke detector within 7.5 m of the door of every habitable room;

(b) detectors designed to be ceiling-mounted should be mounted at least 300 mm from walls and light fittings (unless there is test evidence to demonstrate that such a proximity to light fittings will not adversely affect the functioning of the detector);

(c) the sensor in ceiling-mounted smoke detector/sounder units should be between 25 mm and 600 mm below the ceiling; and

(d) the sensor in ceiling-mounted heat detector/sounder units should be between 25 mm and 150 mm below the ceiling.

Detectors designed for wall mounting may also be used if installed above the level of doorways opening into the circulation space and fixed in accordance with manufacturers' instructions.

1.16   (ADB Vol. 1)

Smoke alarm/detector units should be sited in easily accessible positions so that routine maintenance activities such as testing and cleaning can be carried out safely. They should not be placed over stairs or any openings between floors.

The guidance given in Sections 1.3 to 1.6 in ADB Vol. 1 and above is in line with the guidance and recommendations given in BS 5839-6:2013 Clauses 11.1.1 and 11.2.

1.17 (ADB Vol. 1)

In order to minimise the likelihood of false alarms, smoke detectors/alarms should not be installed:

(a) next to or directly over heaters or air-conditioning units/outlets;
(b) in bathrooms;
(c) in showers;
(d) in kitchens/cooking areas;
(e) in garages; and
(f) in locations where steam, condensation or fumes are likely to be present.

1.18 (ADB Vol. 1)

Similarly, smoke detectors/alarms should not be installed in locations that:

(a) get very hot, such as boiler rooms and laundry areas; and
(b) can be very cold, such as unheated porches.

Fixing smoke detectors/alarms to surfaces which are likely to be considerably warmer or colder than their immediate surroundings should be avoided as this would present the possibility of air currents developing which may prevent or delay smoke from entering the detector unit.

The guidance given in Sections 1.17 and 1.18 in ADB Vol. 1 and above is in line with the guidance and recommendations given in BS 5839-6:2013 Clauses 12.1 and 12.2.

1.19 (ADB Vol. 1)

The power supply to the smoke detector/alarm system should come from either an independent circuit originating from the main distribution board or a regularly used local lighting circuit. In the case of radio-linked units, refer to Section 1.21 in ADB Vol. 1.

Where a lighting circuit is used as the source of supply, a means to isolate the supply to the smoke detector/alarm system and not the lighting should be provided.

The logic behind permitting the supply to be taken from a local lighting circuit is that users of the property will become aware within a very short period of time if the supply is lost for any reason.

1.20 (ADB Vol. 1)

Any electrical installation work in domestic dwellinghouses should comply with Approved Document P (Electrical safety).

1.21 (ADB Vol. 1)

In general, any cable that can be used for the wiring of domestic premises may be used for the power supply for and interconnection between smoke detector/alarm units. However, in large dwellinghouses (as defined above in 1.5) BS 5839-6:2013 specifies the use of fire-resisting cables for Grade A and B systems.

Any cable used for the interconnection of smoke detector/alarm units should be readily distinguishable from cables forming part of the general low voltage electrical installation within the premises, typically by colour coding.

The interconnection between mains-powered smoke detector/alarm units may be via radio-link so long as this does not reduce the standby capacity of the units to below 72 hours' duration.

It is acceptable for units that are radio-linked to be supplied from a number of separate circuits.

1.22   (ADB Vol. 1)

Other options for power supplies are permitted by BS 5839-1:2017 and BS 5839-6:2013 and reference should be made to these standards where it is intended to use measures other than those described in 1.19 and 1.21.

1.23   (ADB Vol. 1)

Fire detection and alarm systems should be properly designed, installed and maintained.

Upon completion of the installation work, installation and completion certificates based on the model forms in Annexes E and F of BS 5839-1:2017 as appropriate for the grade of system installed should be provided.

It is also necessary for the installation of cables and wiring used for the power supply to, and interconnection of, smoke detector/alarm units to be inspected and tested in accordance with the requirements of BS 7671, and for appropriate certification based on the model forms contained in Appendix 6 of that standard to be provided.

1.24   (ADB Vol. 1)

Any and all relevant information relating to the use and maintenance of the alarm system and its component parts should be passed on to the occupants and users of the protected premises.

For further guidance on the information required to be passed on to occupants and users, reference should be made to the following as appropriate for the grade of system installed:

▶ BS 5839-1:2017 Section 40 *Documentation*
▶ BS 5839-6:2013, Section 24 *User instructions*.

Volume 2: Buildings other than dwellinghouses

Note: Volume 2 is relevant to buildings other than dwellinghouses. For guidance applicable to dwellinghouses and the individual dwellings forming part of controlled sheltered housing, reference should be made to Volume 1.

1.2   (ADB Vol. 2)

Sections 1.3, 1.4 and 1.5 in ADB Vol. 2 and below relate to fire alarm and fire detection systems in flats.

1.3   (ADB Vol. 2)

It is recognised that the installation of an automatic fire detection and alarm system can significantly increase the level of safety in such premises by raising the alarm automatically in the event of a fire breaking out. The provisions set out in 1.4 and 1.5 may be considered to be minimum requirements. In the case of flats used to house persons considered to be at an increased risk in the event of a fire, a higher standard of protection may need to be provided.

1.4   (ADB Vol. 2)

A fire detection and alarm system should be installed in all new flats which meets, as a minimum, the recommendations of BS 5839-6:2013 for a Grade D, Category LD3 system. BS 5839-6:2013 defines a Grade D system as 'A system of one or more mains-powered smoke alarms, each with an integral standby supply.'

A Category LD3 system is defined in BS 5839-6:2013 as 'A system incorporating detectors in all circulation spaces that form part of the escape routes from the dwelling.'

1.5   (ADB Vol. 2)

Any smoke and heat alarms used should be mains-operated and have a standby supply from a battery or capacitor.

Smoke alarms should conform to BS EN 14604:2005 *Smoke alarm devices.*

Heat alarms should conform to BS 5446-2:2003 *Fire detection and fire alarm devices – Specification for heat alarms.*

1.6   (ADB Vol. 2)

Where alterations or additions are made such that new habitable rooms are provided above ground floor level, or at ground floor level and there is no final exit from these newly provided rooms, it will be necessary to install a fire detection and alarm system. Smoke alarms will be required in the circulation spaces in accordance with Sections 1.10 to 1.18.

The intention is that occupants of these new rooms will receive warning of a fire occurring in areas forming part of their escape route from the premises.

1.7   (ADB Vol. 2)

The fire detection and alarm systems in the individual dwellings in a sheltered housing scheme under the control of a warden or supervisor should be connected to a central monitoring point or alarm receiving centre. It should be possible for the person watching over the monitoring point for the sheltered housing complex to identify in which individual dwelling(s) a fire alarm has been raised.

It should be remembered that communal areas of a sheltered housing scheme should also meet the recommendations of Volume 2 of Approved Document B, as should the sheltered accommodation of institutional/residential premises such as nursing accommodation, police section houses and the like. The type of system to be installed in such premises should be considered on a case-by-case basis. The guidance given in Sections 1.24 to 1.38 in ADB Vol. 2 and described below should be taken into consideration.

1.8   (ADB Vol. 2)

In situations where up to six students share a self-contained flat having its own entrance door and where the flat has been constructed following the compartmentation principles for flats given in Section 7 of Approved Document B3: *Internal fire spread (structure)*, each flat may be provided with a separate automatic fire detection and alarm system.

In situations such as halls of residence where a general evacuation will be required, a fire detection/alarm system should be installed that follows the guidance relating to buildings other than flats given in Sections 1.24 to 1.38 in ADB Vol. 2.

1.9   (ADB Vol. 2)

In general the design and installation of the fire detection and alarm systems in flats should be in accordance with the recommendations given in BS 5839-6:2013. However, the following points are emphasised in ADB Volume 2.

1.10   (ADB Vol. 2)

In order that an alarm is raised in the early stages of a fire, smoke alarms should be installed in circulation spaces such as corridors, hallways or landings between bedrooms, and those places where fires are most likely to start (that is, kitchens and living rooms).

1.11   (ADB Vol. 2)

At least one smoke detector should be installed on every storey.

1.12   (ADB Vol. 2)

An open-plan situation where the kitchen area is not separated from stairways or circulation spaces by a door requires a compatible heat detector or heat alarm (in terms of the system to which it is to be connected) to be placed in the kitchen area, which should be interconnected to the smoke detector(s) installed in the circulation areas.

1.13   (ADB Vol. 2)

Where the alarm system consists of more than one detector/sounder unit, they should be linked so that detection of smoke or heat by one unit raises the alarm in all of them simultaneously. Any manufacturers' recommendations regarding the maximum number of units that can be linked should be observed.

1.14   (ADB Vol. 2)

The following should be observed when siting detectors on typical flat ceilings:

(a) in a circulation area there should be a smoke detector within 7.5 m of the door of every habitable room;
(b) detectors designed to be ceiling mounted should be mounted at least 300 mm from walls and light fittings (unless there is test evidence to demonstrate that such a proximity to light fittings will not adversely affect the functioning of the detector);
(c) the sensor in ceiling mounted smoke detector/sounder units should be between 25 mm and 600 mm below the ceiling; and
(d) the sensor in ceiling mounted heat detector/sounder units should be between 25 mm and 150 mm below the ceiling.

Detectors designed for wall mounting may also be used if installed above the level of doorways opening into the circulation space and fixed in accordance with manufacturers' instructions.

1.15 (ADB Vol. 2)

Smoke alarm/detector units should be sited in easily accessible positions so that routine maintenance activities such as testing and cleaning can be carried out safely. They should not be placed over stairs or any openings between floors.

1.16 (ADB Vol. 2)

In order to minimise the likelihood of false alarms, smoke detectors/alarms should not be installed:

(a) next to or directly over heaters or air-conditioning units/outlets;
(b) in bathrooms;
(c) in showers;
(d) in kitchens/cooking areas;
(e) in garages; and
(f) in locations where steam, condensation or fumes are likely to be present.

1.17 (ADB Vol. 2)

Similarly, smoke detectors/alarms should not be installed in locations that:

(a) get very hot, such as boiler rooms and laundry areas; and
(b) can be very cold, such as unheated porches.

Fixing smoke detectors/alarms to surfaces which are likely to be considerably warmer or colder than their immediate surroundings should be avoided as this would present the possibility of air currents developing which may prevent or delay smoke from entering the detector unit.

The guidance given in Sections 1.16 and 1.17 in ADB Vol. 1 and above is in line with the guidance and recommendations given in BS 5839-6 Clauses 12.1 and 12.2.

1.18 (ADB Vol. 2)

Any and all relevant information relating to the use and maintenance of the alarm system and its component parts should be passed on to the occupants and users of the protected premises.

For further guidance on the information required to be passed on to occupants and users, reference should be made to the following as appropriate for the grade of system installed:

1   BS 5839-1:2017, Section 40 *Documentation*
2   BS 5839-6:2013, Section 24 *User instructions*.

1.19 (ADB Vol. 2)

The power supply to the smoke detector/alarm system should come from either an independent circuit originating from the main distribution board or a regularly used local lighting circuit. In the case of radio-linked units, refer to Section 1.21 in ADB Vol. 1.

Where a lighting circuit is used as the source of supply, a means to isolate the supply to the smoke detector/alarm system and not the lighting should be provided.

The logic behind permitting the supply to be taken from a local lighting circuit is that users of the property will become aware within a very short period of time if the supply is lost for any reason.

1.20   (ADB Vol. 2)

The electrical installation should comply with Approved Document P (Electrical safety).

1.21   (ADB Vol. 2)

In general, any cable suitable for the wiring of domestic premises may be used for the power supply for and interconnection between smoke detector/alarm units.

Any cable used for the interconnection of smoke detector/alarm units should be readily distinguishable from cables forming part of the general low voltage electrical installation within the premises, typically by colour coding.

The interconnection between mains-powered smoke detector/alarm units may be via radio-link provided that this does not reduce the standby capacity of the units to below 72 hours' duration.

It is acceptable for units that are radio-linked to be supplied from a number of separate circuits.

1.22   (ADB Vol. 2)

Other options for power supplies are permitted by BS 5839-1:2017 and BS 5839-6:2013 and reference should be made to these standards where it is intended to use measures other than those described in 1.19 and 1.21.

1.23   (ADB Vol. 2)

Fire detection and alarm systems should be properly designed, installed and maintained.

Upon completion of the installation work, installation and completion certificates based on the model forms in BS 5839 as appropriate for the grade of system installed should be provided.

It is also necessary for the installation of cables and wiring used for the power supply to, and interconnection of, smoke detector/alarm units to be inspected and tested in accordance with the requirements of BS 7671 and for appropriate certification based on the model forms contained in Appendix 6 of that standard to be provided.

1.24   (ADB Vol. 2)

When deciding which type of fire detection/alarm system needs to be installed, consideration should be given to pertinent factors such as type of occupancy and the escape strategy that is to be employed (simultaneous, phased, or progressive horizontal evacuation).

1.25   (ADB Vol. 2)

The threat posed by fire is considered to be much greater in residential accommodation where persons are expected to sleep than in premises where the occupants are expected to be awake and alert.

Where the means of escape is based on simultaneous evacuation, operation of a manual call point ('break-glass' unit) or a detector should result in an almost instantaneous warning being raised by the sounders and other devices as appropriate.

Where the means of escape is based on phased evacuation, a staged alarm response is appropriate, for example, typical alarm stages might be 'alert' and 'evacuate'.

1.26   (ADB Vol. 2)

The type and category of fire detection/alarm system required in any given premises should be decided upon on a case-by-case basis.

General guidance on which type and category of system might be appropriate can be found in Table A1 of BS 5839-1:2017.

1.27   (ADB Vol. 2)

Arrangements should be put in place for detecting fire in all buildings. However, this does not necessarily mean the installation of an automatic fire detection/alarm system.

1.28   (ADB Vol. 2)

It is not a requirement to have an automatic fire detection/alarm system in a small non- domestic building or premises. If, as a result of carrying out a risk assessment, it can be confirmed that a suitable audible fire alarm warning could be raised by, for example, shouting, then this may be considered to be adequate. Other alternatives for raising the alarm that may prove suitable in small non-domestic locations include hand-held bells and other manually operated sounders, or an alarm system consisting only of manual call points and warning devices (that is, containing no automatic detectors) classed as a Category M system in BS 5839-1:2017. See also item 1.30.

1.29   (ADB Vol. 2)

In larger premises not covered by 1.28 in ADB Vol. 2 and above, a building should be provided with a suitable electrically operated fire detection/alarm system containing a sufficient number of detectors and manual call points as required to give adequate coverage of the premises, and sufficient sounders and other warning devices to provide an adequate notification of the alarm situation throughout the premises.

1.30   (ADB Vol. 2)

Any electrically operated fire detection/alarm system that is installed in non-domestic premises covered by ADB Vol. 2 should comply with all relevant parts of BS 5839-1:2017.

BS 5839-1:2017 recognises three categories of system:

(a) Category L – offering life protection (for human occupants of premises)
(b) Category P – offering property protection for the premises
(c) Category M – manually operated fire alarm systems.

For further information on Category M classification refer to Clause 5.1.2 of BS 5939-1:2017.

Categories L and P are further subdivided as follows:

(a) *Category L1*A life protection system giving coverage throughout all relevant parts of the building.

(b) *Category L2*A life protection system giving coverage only to defined parts of the protected building. Note that a Category L2 system should include the coverage required of a Category L3 system as described below.

(c) *Category L3*A life protection system designed to give a warning of a fire sufficiently early to enable all occupants of a protected building, other than possibly those persons in the room where the fire originally occurred, to escape safely, before escape routes become impassable due to the presence of smoke, flames, toxic gases and other products of combustion.

(d) *Category L4*A life protection system giving coverage of those parts of the escape routes comprising circulation areas and circulation spaces including corridors and stairways. The purpose of such a system is to enhance the safety of occupants by giving them early warning of the presence of smoke within escape routes through which they will need to travel to reach a place of safety.

(e) *Category L5*A life protection system in which the protected area(s) and/or the location of the detectors is designed to meet a specific fire safety objective somewhat different to that of the Category L1, L2, L3 or L4 systems described above.

For further information on Category L classifications refer to Clause 5.1.3 of BS 5839-1:2017.

(a) *Category P1*A property protection system installed throughout the protected building.

(b) *Category P2*A property protection system installed only in defined parts of the building. The purpose of such a system is to allow a fire alarm to be raised as soon as possible in areas which present a high fire hazard, or in which the risk to property or business continuity is high from a fire.

For further information on Category P classifications refer to Clause 5.1.4 of BS 5839-1:2017.

1.31   (ADB Vol. 2)

Any manual call points used as part of a fire detection/alarm system should be of Type A (direct) operation as given in BS EN 54-11. A Type A call point is defined as 'a manual call point in which the change to the alarm condition is automatic (i.e. without the need for further manual action) when the frangible element is broken or displaced'. The manual call points should be installed in accordance with the relevant Clauses of BS 5839-1:2017.

Type B operation manual call points as given in BS EN 54-11:2001, having indirect operation where, for example, it is necessary to break the glass and then press a button to raise the alarm, may only used with the approval of the relevant local Building Control body.

1.32   (ADB Vol. 2)

In situations where it is believed that persons may not respond quickly to a fire warning, or where persons are unfamiliar with the fire warning arrangements, consideration may be given to the use of a voice alarm system. Such a system may form part of a general public address system within the building and could give both an audible signal and verbal instructions for occupants to follow in the event of a fire.

Any such warning signal should be recognisably different and distinct from any other signals in general use. Any accompanying verbal instructions should be clear.

Any installed voice alarm system should comply with the relevant Clauses of BS 5839-8:2013 *Fire detection and fire alarm systems for buildings. Code of practice for the design, installation, commissioning and maintenance of voice alarm systems.*

1.33 (ADB Vol. 2)

It may not be desirable to raise a general alarm in large premises such as shopping centres and places of assembly as a result of the large number of members of the general public that are likely to be present at any given time. In such instances where a general alarm is not raised, it is important that sufficient numbers of appropriately trained staff are in place to initiate any pre-planned procedures for safe evacuation. Actuation of the fire alarm system will alert staff by means of discreet sounders, personal messages or paging or similar means.

Where full evacuation of such premises is necessary, notification can be given by sounders or a message broadcast over the public address system. The installed fire alarm system should in all other respects comply with BS 5839-1:2017.

1.34 (ADB Vol. 2)

In situations where it is reasonably likely that one or more persons having impaired hearing are likely to be present in relative isolation such as in a hotel room whilst staying as a guest, or whilst using lavatories or in other sanitary accommodation and where no other method of alerting them is available, the audible fire alarm signal should be supplemented by a visual indication.

In buildings such as schools, colleges and offices where the persons present are under some control, the use of a vibrating pager system or similar might be appropriate.

Reference should be made to BS 5839-1:2017, Clause 18, 'Fire alarm warnings for people with impaired hearing'.

1.35 (ADB Vol. 2)

An automatic fire detection/alarm system in accordance with BS 5839-1:2017 should be provided in all institutional and residential premises where it is expected that persons will sleep. This would include, for example, lodging blocks for students, nurses, police and the like.

1.36 (ADB Vol. 2)

Although automatic fire detection systems are not generally required in non-residential premises/buildings, there may be specific circumstances making the installation of such a BS 5839-1:2017 type system necessary, such as:

1 to compensate for one or more departures from the guidance on fire safety given elsewhere in ADB;
2 to act as part of the operating system for a greater fire protection system incorporating, for example, pressure differential systems or automatic door release mechanisms; and

3   where there exists the possibility of a fire breaking out in a part of the premises which could have an adverse effect upon the means of escape from those premises. This could include locations not frequently visited in normal use such as storage areas, and parts of a building that have been vacated and are not currently in regular use.

1.37   (ADB Vol. 2)

Any installed fire detection/alarm system must be properly designed, installed and maintained. On completion, or as soon as possible after completion, of the design, installation and commissioning of the system, the appropriate certification should be completed by the person responsible for that aspect of the fire alarm system.

1.38   (ADB Vol. 2)

Fire detection/alarm systems can form a part of a larger fire protection regime within a building and may be used to initiate the operation, or change of state of other constituent parts of such a greater system such as smoke control systems; fire extinguishing systems; door release mechanisms for fire doors; automatic unlocking arrangements of exit doors, and the like.

As such, any interfaces between the fire detection/alarm system and other systems required to achieve compliance with Building Regulations must be designed to give a high degree of reliability of operation.

Particular care must be exercised if the interface is made via another system such as an access control system.

Where any part of BS 7273 *Code of practice for the operation of fire protection measures* applies to the actuation of other systems such as gaseous total flooding systems (Parts 1 and 2), release mechanisms for doors (Part 4) or non pre-action water mist systems (Part 5), the recommendations of the relevant part of that standard should be followed.

Appendix C

Approved Document B2: Internal fire spread (linings)

For Scotland see Appendix E of this Guidance Note.

For Northern Ireland see the Technical Booklets listed in Explanatory Note 8 to the Building Regulations (Northern Ireland) 2012 as amended. The publications include Technical Booklet E:2012 and other fire safety documents. The legislation and Technical Booklet can be found on the Department of Finance and Personnel's building regulations website: www.buildingregulationsni.gov.uk

For England and Wales, Approved Document B (Fire safety) is subdivided into two volumes. Volume 1 contains guidance applicable to dwellinghouses while Volume 2 contains guidance applicable to buildings other than dwellinghouses. Consequently the guidance in this Appendix is similarly divided along these lines.

In the text below, Approved Document B is abbreviated to ADB.

At the date of printing of this Guidance Note, different versions of ADB apply in England and Wales, as follows:

(a) England:
- **(i)** Volume 1 – 2006 edition incorporating 2010, 2013 and 2017 amendments
- **(ii)** Volume 2 – 2006 edition incorporating 2007, 2010, 2013 and 2017 amendments

(b) Wales:
- **(i)** Volume 1 – 2006 edition incorporating 2010 and 2017 amendments
- **(ii)** Volume 2 – 2006 edition incorporating 2010, 2013 and 2017 amendments

Unless otherwise indicated the text in this appendix relates to both England and Wales.

Volume 1: Dwellinghouses

Note: Volume 1 is relevant to houses and the individual residential units of sheltered housing. For guidance applicable to the common parts of a sheltered housing development, flats, student accommodation and other buildings, reference should be made to Volume 2.

Volume 2: Buildings other than dwellinghouses

Note: Volume 2 is relevant to buildings other than dwellinghouses. For guidance applicable to dwellinghouses and the individual dwellings forming part of controlled sheltered housing, reference should be made to Volume 1.

The provisions described in Vol. 1 Sections 3.11-3.13, Vol. 2 England Sections 6.13-6.15, and Vol. 2 Wales Sections 7.13-7.15 apply to lighting diffusers which form part of a ceiling. They do not relate to diffusers of light fittings which are attached to the soffit of, or suspended beneath, a ceiling. See Figure C.1.

▼ **Figure C.1** Lighting diffuser in relation to ceiling

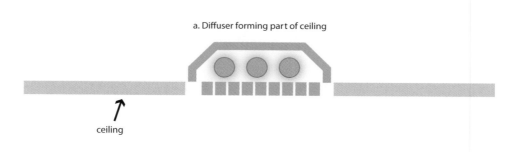

a. Diffuser forming part of ceiling

ceiling

b. Diffuser in fitting attached below and not forming part of ceiling

ceiling

Thermoplastic lighting diffusers should not be used in fire-protecting or fire-resisting ceilings, unless they have been satisfactorily tested as part of a ceiling system that can be used to provide the appropriate fire protection.

Subject to ADB Vol. 1 Sections 3.11-3.13, Vol. 2 England Sections 6.13-6.15, and Vol. 2 Wales Sections 7.13-7.15, ceilings of rooms and circulation spaces (but **not** protected stairways) may contain thermoplastic lighting diffusers if the following provisions are followed:

(a) The exposed wall and ceiling surfaces above the suspended ceiling (other than the upper surfaces of the thermoplastic panels themselves) should comply with the general provisions of Section 3.1 of ADB Vol. 1 and Section 6.1 of ADB Vol. 2 (Classification of linings)

(b) For diffusers of classification TP(a) (rigid), no restrictions on their use apply

(c) For diffusers of classification TP(b), their use should be restricted in extent as given in Figure C.2 and Table C.1.

▼ **Figure C.2** Layout restrictions on Class 3 plastic rooflights, TP(b) rooflights and TP(b) lighting diffusers

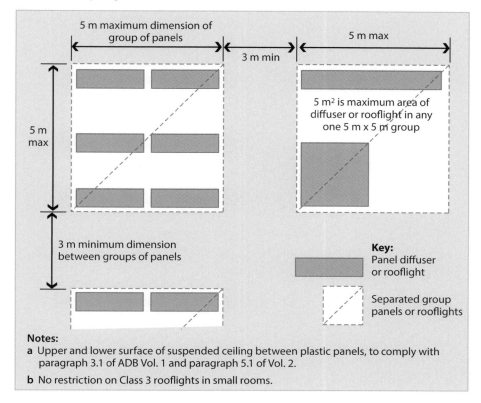

Notes:

a Upper and lower surface of suspended ceiling between plastic panels, to comply with paragraph 3.1 of ADB Vol. 1 and paragraph 5.1 of Vol. 2.

b No restriction on Class 3 rooflights in small rooms.

▼ **Table C.1** Classification of diffusers

Minimum classification of lower surface	Use of space below the diffusers or rooflight	Maximum area of each diffuser panel or rooflight[1]	Maximum total area of diffuser panels and rooflights as percentage of floor area of the space in which the ceiling is located	Minimum separation distance between diffuser panels or rooflights[1]
TP(a)	Any except protected stairway	No limit[2]	No limit	No limit
Class 3[3] or TP(b)	Rooms	5 m²	50 %[4]	3 m
	Circulation spaces except protected stairways	5 m²	15 %[4]	3 m

Notes:1 Smaller panels can be grouped together provided that the overall size of the group and the space between one group and any others satisfies the dimensions shown in Figure C.2.

2 Lighting diffusers of TP(a) flexible rating should be restricted to panels of not more than 5 m² each, see Section 3.14.

3 There are no limitations on Class 3 material in small rooms.

4 The minimum 3 m separation specified in Figure C.2 between each 5 m² must be maintained. Therefore, in some cases it may not be possible to use the maximum percentage quoted.

Appendix

Approved Document B3:
Internal fire spread (structure)

D

For Scotland see Appendix E of this Guidance Note.

For Northern Ireland see the Technical Booklets listed in Explanatory Note 8 to the Building Regulations (Northern Ireland) 2012 as amended. The publications include Technical Booklet E:2012 and other fire safety documents. The legislation and Technical Booklet can be found on the Department of Finance and Personnel's building regulations website: www.buildingregulationsni.gov.uk

For England and Wales, Approved Document B (Fire safety) is subdivided into two volumes. Volume 1 contains guidance applicable to dwellinghouses while Volume 2 contains guidance applicable to buildings other than dwellinghouses. Consequently the guidance in this Appendix is similarly divided along these lines.

In the text below, Approved Document B is abbreviated to ADB.

At the date of printing of this Guidance Note, different versions of ADB apply in England and Wales, as follows:

(a) England:
   (i) Volume 1 – 2006 edition incorporating 2010, 2013 and 2017 amendments
   (ii) Volume 2 – 2006 edition incorporating 2007, 2010, 2013 and 2017 amendments

(b) Wales:
   (i) Volume 1 – 2006 edition incorporating 2010 and 2017 amendments
   (ii) Volume 2 – 2006 edition incorporating 2010, 2013 and 2017 amendments

Unless otherwise indicated the text in this appendix relates to both England and Wales.

Volume 1: Dwellinghouses

Note: Volume 1 is relevant to houses and the individual residential units of sheltered housing. For guidance applicable to the common parts of a sheltered housing development, flats, student accommodation and other buildings, reference should be made to Volume 2.

Volume 2: Buildings other than dwellinghouses

Note: Volume 2 is relevant to buildings other than dwellinghouses. For guidance applicable to dwellinghouses and the individual dwellings forming part of controlled sheltered housing, reference should be made to Volume 1.

7.2   (Vol. 1), 10.2 (Vol. 2 England) and 11.2 (Vol. 2 Wales)

If a fire-separating element of construction such as a ceiling, floor, wall or similar is to be effective, it is essential that every joint, imperfection of fit or, of particular interest to this Guidance Note, opening to allow services to pass through such elements should be adequately sealed or fire-stopped such that the fire resistance of the element is not reduced.

7.3   (Vol. 1), 10.3 (Vol. 2 England) and 11.3 (Vol. 2 Wales)

The measures described in Section 7 of ADB Vol. 1, Section 10 of ADB Vol. 2 England and Section 11 of ADB Vol. 2 Wales, 'Protection of openings and fire-stopping' are primarily intended to delay the passage of fire through a building. It is accepted that these measures will additionally retard smoke spread through a building. However, there are no specific criteria relating to passage of smoke.

7.4   (Vol. 1)

Consideration should be given to the effect that the installation of electrical equipment and accessories including downlighters may have in reducing the fire resistance of elements of the construction. In some cases some additional protective measures may be required.

7.12   (Vol. 1), 10.17 (Vol. 2 England) and 11.17 (Vol. 2 Wales)

Fire-stopping is required in all

(a) joints between fire-separating elements of the construction; and
(b) openings made to allow pipes, ducts, conduits, trunking or cables to pass through fire-separating elements of the construction.

Any such openings for building services should be as physically small as possible and their numbers kept to a minimum.

7.13   (Vol. 1), 10.18 (Vol. 2 England) and 11.18 (Vol. 2 Wales)

In order to prevent displacement and hence reduced effectiveness, any materials used for fire-stopping should be reinforced with, or be supported by, materials having limited combustibility:

(a) where the unsupported span is greater than 100 mm; and
(b) in any other case where non-rigid materials are used (unless it has been demonstrated by test that they are suitable without such reinforcement).

7.14   (Vol. 1), 10.19 (Vol. 2 England) and 11.19 (Vol. 2 Wales)

A number of proprietary fire-stopping and sealing products, some of which are specifically designed for service penetrations, that have been shown by test to maintain fire resistance of construction elements are available. These may be used in accordance with manufacturers' instructions.

Other materials which may be employed for fire-stopping and sealing include:

(a) cement-based or gypsum-based vermiculite/perlite mixes;
(b) cement mortar;
(c) glass fibre-, crushed rock-, blast furnace slag-, or ceramic-based products (with or without additional binding elements);

(d) gypsum-based plaster; and

(e) intumescent mastics.

It should be recognised that not all of the above will be suitable for all situations and an appropriate material should be selected to suit particular individual circumstances.

Reference should also be made to Section 6.2 of this Guidance Note.

D

Guidance Note 4: Protection Against Fire
© The Institution of Engineering and Technology

Appendix E

Scottish Technical Standards relating to protection against fire

For England and Wales see Appendices B, C and D of this Guidance Note.

For Northern Ireland see the Technical Booklets listed in Explanatory Note 8 to the Building Regulations (Northern Ireland) 2012 as amended. The publications include Technical Booklet E:2012 and other fire safety documents. The legislation and Technical Booklet can be found on the Department of Finance and Personnel's building regulations website: www.buildingregulationsni.gov.uk

The Scottish Government Building Standards Division (BSD) has produced two Technical Handbooks intended to provide guidance on achieving the standards set in the Building (Scotland) Regulations 2004 (as amended). One Technical Handbook relates to Domestic buildings and the other to Non-domestic buildings.

Compliance with the Building (Scotland) Regulations 2004 can be achieved by compliance with the mandatory Building Standards and the associated guidance given in the Technical Handbooks.

The last revision of the Technical Handbooks came into force on 28 June 2017.

The information contained in this appendix relates to Section 2 'Fire' and Section 4 'Safety' of the Technical Handbooks.

An indication of whether the mandatory Building Standard is applicable to domestic and/or non-domestic buildings is given in the margin.

Where dated versions of British Standards are referred to in this appendix, these versions are the ones referred to in the Technical Booklets, which is some cases are not the latest version of the British Standard concerned. Where this is the case, compliance with the relevant provisions of the Building Regulations will generally be achieved by following the corresponding provisions of the latest version of the British Standard concerned.

2.1   Compartmentation

Non-domestic

Every building must be designed and constructed in such a way that in the event of an outbreak of fire within the building, fire and smoke are inhibited from spreading beyond the compartment of origin until any occupants have had the time to leave that compartment and any fire containment measures have been initiated.

Limitation:

This standard does not apply to domestic buildings.

2.1.14 Non-domestic

This section deals with openings and service penetrations.

Any openings or service penetrations in elements of the building structure should be kept to a minimum.

Self-closing fire doors can be fitted with hold open devices as specified in BS 5839-3:1988 *Fire detection and fire alarm systems for buildings. Specification for automatic release mechanisms for certain fire protection equipment* provided that the door is not an emergency door, a protected door serving the only escape stair in the building (or the only escape stair serving part of the building) or a protected door serving a fire-fighting shaft.

If it is decided to fit automatic release mechanisms as described above, this may make it necessary to install an automatic fire detection/alarm system.

Any service opening other than a ventilating duct which penetrates a compartment wall or compartment floor should be fire-stopped such that it provides at least the same duration of fire resistance as if the wall or floor had not been so penetrated. This may be achieved by:

(a) a casing which has at least the appropriate fire resistance from the outside; or
(b) a casing which has at least half the appropriate fire resistance from each side; or
(c) an automatic heat-activated sealing device that will maintain the appropriate fire resistance in respect of integrity for the wall or floor regardless of the opening size.

Fire-stopping need not be provided for:

(a) a pipe or a cable with a bore, or diameter, of not more than 40 mm; or
(b) not more than four 40 mm diameter pipes or cables that are at least 40 mm apart and at least 100 mm from any other pipe; or
(c) more than four 40 mm diameter pipes or cables that are at least 100 mm apart.

Fire-stopping may be required to close an imperfection of fit or design tolerance between construction elements and components, around service openings and ventilation ducts. Any proprietary fire-stopping products, including intumescent products, should be subjected to testing to demonstrate their ability to maintain a sufficient degree of fire resistance under the conditions in which they are to be employed.

Where only minimal differential movement is anticipated, either in normal use or during fire exposure, the following may be used:

(a) proprietary fire-stopping products;
(b) cement mortar;
(c) gypsum-based plaster;
(d) cement- or gypsum-based vermiculite/perlite mixes; and
(e) mineral fibre, crushed rock and blast furnace slag or ceramic-based products (with or without added binders).

Where greater differential movement is anticipated, either in normal use or during fire exposure, proprietary fire-stopping products may be used.

In order to prevent their displacement, materials used for fire-stopping should be provided with reinforcement or support from non-combustible materials where the unsupported span is more than 100 mm and where non-rigid materials are used. This will not be necessary where it has been shown by test that the materials alone are satisfactory within their field of application.

Reference should also be made to Section 6.2 of this Guidance Note.

2.2   Separation

Domestic and non-domestic | Every building that is divided into more than one area of different occupation must be designed and constructed in such a way that in the event of an outbreak of fire within the building, fire and smoke are inhibited from spreading beyond the area of occupation where the fire originated.

2.2.4   Non-domestic and 2.2.6 Domestic

In order to reduce the risk of a fire starting within a combustible separating wall or a fire spreading rapidly on or within the wall construction:

(a) insulation material exposed in a cavity should be of low risk or non-combustible materials (see Annex 2.E, Domestic, or 2.B, Non-domestic); and

(b) the internal wall lining should be constructed from material which is low risk or non-combustible; and

(c) the wall should contain no pipes, wires or other services.

2.2.9   Openings and service penetrations – domestic

Any openings or service penetrations in elements of the building structure should be kept to a minimum.

Self-closing fire doors can be fitted with hold open devices as specified in BS 5839-3:1988 *Fire detection and fire alarm systems for buildings. Specification for automatic release mechanisms for certain fire protection equipment* provided that the door is not an emergency door, a protected door serving the only escape stair in the building (or the only escape stair serving part of the building) or a protected door serving a fire-fighting shaft.

If it is decided to fit automatic release mechanisms as described above, this may make it necessary to install an automatic fire detection/alarm system.

Any service opening other than a ventilating duct which penetrates a separating wall or separating floor should be fire-stopped such that it provides at least the same duration of fire resistance as if the wall or floor had not been so penetrated. This may be achieved by:

(a) a casing which has at least the appropriate fire resistance from the outside; or

(b) a casing which has at least half the appropriate fire resistance from each side; or

(c) an automatic heat-activated sealing device that will maintain the appropriate fire resistance in respect of integrity for the wall or floor regardless of the opening size.

Fire stopping need not be provided for:

(a) a pipe or a cable with a bore, or diameter, of not more than 40 mm; or
(b) not more than four 40 mm diameter pipes or cables that are at least 40 mm apart and at least 100 mm from any other pipe; or
(c) more than four 40 mm diameter pipes or cables that are at least 100 mm apart.

Fire-stopping may be required to close an imperfection of fit or design tolerance between construction elements and components, around service openings and ventilation ducts. Any proprietary fire-stopping products, including intumescent products, should be subjected to testing to demonstrate their ability to maintain a sufficient degree of fire resistance under the conditions in which they are to be employed.

Where only minimal differential movement is anticipated, either in normal use or during fire exposure, the following may be used:

(a) proprietary fire-stopping products;
(b) cement mortar;
(c) gypsum-based plaster;
(d) cement- or gypsum-based vermiculite/perlite mixes; and
(e) mineral fibre, crushed rock and blast furnace slag or ceramic-based products (with or without added binders).

Where greater differential movement is anticipated, either in normal use or during fire exposure, proprietary fire-stopping products may be used.

In order to prevent their displacement, materials used for fire-stopping should be provided with reinforcement or support from non-combustible materials where the unsupported span is more than 100 mm and where non-rigid materials are used. This will not be necessary where it has been shown by test that the materials alone are satisfactory within their field of application.

Reference should also be made to Section 6.2 of this Guidance Note.

2.3   Structural protection

Domestic and non-domestic

Every building must be designed and constructed in such a way that in the event of an outbreak of fire within the building, the load-bearing capacity of the building will continue to function until all occupants have escaped, or been assisted to escape, from the building and any fire containment measures have been initiated.

2.3.4   Domestic and non-domestic

In general, openings and service penetrations in elements of structure do not need to be protected from fire unless there is the possibility of structural failure.

Where a large opening or a large number of small openings are made, care should be taken to ensure that the load-bearing capacity of the structure is maintained.

2.4 Cavities

Domestic and non-domestic

Every building must be designed and constructed in such a way that in the event of an outbreak of fire within the building, the unseen spread of fire and smoke within concealed spaces in its structure and fabric is inhibited.

2.5 Internal linings

Domestic and non-domestic

Every building must be designed and constructed in such a way that in the event of an outbreak of fire within the building, the development of fire and smoke from the surfaces of walls and ceilings within the area of origin is inhibited.

2.5.7 Domestic and non-domestic

Thermoplastic materials may be used in light fittings with diffusers. Where the lighting diffuser forms an integral part of the ceiling, the size and disposition of the lighting diffusers should be installed in accordance with the recommendations given in Table E.1 and Figure E.1.

▼ **Table E.1** Thermoplastic rooflights and light fittings with diffusers

Classification of lower surface	Protected zone or fire-fighting shaft	Unprotected zone or protected enclosure		Room	
	Any thermoplastic	TP(a) rigid	TP(a) flexible and TP(b)	TP(a) rigid	TP(a) flexible and TP(b)
Maximum area of each diffuser panel or rooflight	Not advised	No limit	5 m²	No limit	5 m²
Maximum total area of diffuser panels or rooflights as a percentage of the floor area of the space in which the ceiling is located	Not advised	No limit	15 %	No limit	50 %
Minimum separation distance between diffuser panels or rooflights	Not advised	No limit	3 m	No limit	3 m

Notes: (a) Smaller panels can be grouped together provided that the overall size of the group, and the space between any others, satisfies the dimensions shown in Figure E.1.

(b) The minimum 3 m separation in Figure E.1 should be maintained between each 5 m² panel. In some cases, therefore, it may not be possible to use the maximum percentage quoted.

(c) TP(a) flexible is not recommended in rooflights.

Where the lighting diffusers form an integral part of a fire-resisting ceiling which has been satisfactorily tested, the amount of thermoplastic material is unlimited.

Where light fittings with thermoplastic diffusers do not form an integral part of the ceiling, the amount of thermoplastic material is unlimited provided the lighting diffuser is designed to fall out of its mounting when softened by heat.

For further information on thermoplastic categories reference should be made to the Scottish Government Building standards (BSD) Technical Handbooks.

▼ **Figure E.1** Layout restrictions on thermoplastic rooflights and light fittings with diffusers

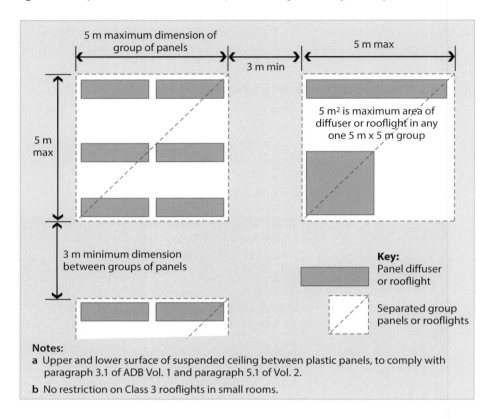

Notes:
a Upper and lower surface of suspended ceiling between plastic panels, to comply with paragraph 3.1 of ADB Vol. 1 and paragraph 5.1 of Vol. 2.

b No restriction on Class 3 rooflights in small rooms.

2.9   Escape

Domestic and non-domestic Every building must be designed and constructed in such a way that in the event of an outbreak of fire within the building, the occupants, once alerted to the outbreak of the fire, are provided with the opportunity to escape from the building, before being affected by fire or smoke.

2.10   Escape lighting

Domestic and non-domestic Every building must be designed and constructed in such a way that in the event of an outbreak of fire within the building, illumination is provided to assist in escape.

2.10.1 Domestic

Escape lighting is not required for dwellings as it is assumed that occupants will be familiar with the layout. In buildings containing flats and maisonettes, any common escape routes will require illumination to assist occupants to find their way to a place of safety.

2.10.2 Non-domestic

Escape routes in non-domestic premises should be illuminated to aid the safe evacuation of a building in an emergency.

It may be necessary to provide emergency lighting and exit signage under the Fire Safety (Scotland) Regulations 2006 (for further information, refer to Clause 2.0.8 of the non-domestic Technical Handbook).

Part 1 of the Cinematograph (Safety) (Scotland) Regulations 1955 contains specific requirements for lighting for buildings such as cinemas.

Additional guidance on special fire precautions that may be required within residential care buildings, hospitals and enclosed shopping centres is grouped in the annexes to the non-domestic Technical Handbook as follows:

(a) residential care buildings, see Annex 2.A
(b) hospitals, see Annex 2.B
(c) enclosed shopping centres, see Annex 2.C.

These annexes should help designers and verifiers to find the information they require quickly when designing or vetting such buildings.

It should be remembered that the guidance in the annexes is in addition and supplementary to the guidance to Standard 2.1 to 2.15.

2.10.3 Escape route lighting – domestic and non-domestic

Escape routes should be provided with artificial lighting supplied by a protected circuit to provide a level of illumination not less than that recommended for emergency lighting (see Clause 2.10.5). Where artificial lighting serves a protected zone, it should be supplied from a protected circuit separate from that supplying any other part of the escape route.

Artificial lighting supplied by a protected circuit need not be provided where a system of emergency lighting is installed.

2.10.4 Protected circuits – domestic and non-domestic

A protected circuit is a circuit originating at the main incoming switch or distribution board, the conductors of which are protected against fire.

Regardless of what system is employed, escape routes should be capable of being illuminated when the building is in use.

2.10.5 Emergency lighting – domestic and non-domestic

Emergency lighting is lighting designed to come into, or remain in, operation automatically in the event of a local and general power failure and should be installed in buildings considered to be at higher risk, such as in high-rise buildings, buildings with basements or in rooms where the number of people is likely to exceed 60.

Emergency lighting should be installed in buildings or parts of a building considered to be at higher risk such as:

(a) a protected zone and an unprotected zone in a building with any storey at a height of more than 18 m;

(b) a room with an occupancy capacity of more than 60, or in the case of an inner room where the combined occupancy capacity of the inner room plus the adjoining room (and any protected zone or unprotected zone serving these rooms) is more than 60;

(c) an underground car park including any protected zone or unprotected zone serving it where less than 30 % of the perimeter of the car park is open to the external air;

(d) a protected zone or unprotected zone serving a basement storey;

(e) a place of special fire risk (other than one requiring access only for the purposes of maintenance) and any protected zone or unprotected zone serving it; and

(f) a protected zone or unprotected zone serving a storey which has at least two storey exits in the following buildings:

    (i) entertainment, assembly, factory, shop, multi-storey storage (Class 1), single-storey storage (Class 1) with a floor area more than 500 m²;

    (ii) a protected zone or unprotected zone serving a storey in a multi-storey non-residential school; and

    (iii) a protected zone or unprotected zone serving any storey in an open-sided car park.

Domestic – The emergency lighting should be installed in accordance with BS 5266: Part 1: 2005 as read in association with BS 5266: Part 7: 1999 (BS EN: 1838: 2013).

Non-domestic – Emergency lighting in places of entertainment, such as cinemas, bingo halls, ballrooms, dance halls and bowling alleys, should be in accordance with BS 5266: Part 1: 2005. Emergency lighting in any other building should be in accordance with BS 5266: Part 1: 2005 as read in association with BS 5266: Part 7: 1999 (BS EN 1838: 2013).

In the case of a building with a smoke and heat exhaust ventilation system, the emergency lighting should be sited below the smoke curtains or installed so that it is not rendered ineffective by smoke-filled reservoirs.

2.11 Communication

Domestic and non-domestic — Every building must be designed and constructed in such a way that in the event of an outbreak of fire within the building, the occupants are alerted to the outbreak of fire.

2.11.1 Domestic

It is accepted that early detection and warning of fire to occupants of domestic premises can play a vital role in increasing their chances of escape. This is particularly important given that the occupants will be asleep some of the time.

2.11.2 Fire detection and fire alarm systems – domestic

Living rooms and kitchens should be fitted with fire detectors. Also, the circulation spaces outside the bedrooms and other rooms or spaces within a dwelling should be protected to give early warning of fire.

Therefore, in order to provide a fire detection and fire alarm system that should alert occupants to the outbreak of fire a Grade D system should be installed in all dwellings, comprising of at least:

(a) one smoke alarm installed in the principal habitable room;

(b) one smoke alarm in every circulation space such as hallways and landings; and

(c) one heat alarm installed in every kitchen.

A principal habitable room is a frequently used room by the occupants of a dwelling for general daytime living purposes.

Every inner room and adjoining access room should be provided with an additional smoke alarm to give the occupants early warning. Where the access room is a kitchen, the type of detector should be carefully considered to reduce the likelihood of false alarms.

Common systems – in a building containing flats or maisonettes, a common fire alarm and detection system that interlinks all dwellings and common spaces is not recommended due to the risk of unwanted false alarms. However in a sheltered housing complex, monitoring equipment is recommended due to the vulnerability of the occupants.

Detailed guidance on fire detection and fire alarm systems in dwellings can be obtained from BS 5839: Part 6:2004.

2.11.3 Choice of fire detector – domestic

False alarms are common in dwellings and may result in the occupants disabling the fire detection and fire alarm system. The most common causes of a false alarm are:

(a) fumes from cooking (including toasting of bread);

(b) steam from bathrooms, shower rooms and kitchens;

(c) tobacco smoke;

(d) dust;

(e) aerosol spray and incense;

(f) candles;

(g) high humidity; and

(h) water ingress.

Consideration should therefore be given to the type of fire detector in order to reduce the amount of unwanted false alarms. There are 4 main types of fire detector used in dwellings: optical smoke alarms, ionisation smoke alarms, multi sensor alarms and heat alarms. These are discussed in '2.11.3 Domestic' to '2.11.6 Domestic', respectively.

2.11.4 Optical smoke alarms – domestic

Optical smoke alarms should conform to BS EN 14604:2005 and operate on the principle of detecting the scattering or absorption of light within the detector chamber. Optical smoke alarms are more sensitive to slow smouldering fires such as fires involving soft furnishings and bedding.

Principal habitable room – the most likely source of fire in a principal habitable room is the careless disposal of smoking materials. Polyurethane foam found in some furnishings may ignite and begin to smoulder producing large particles of smoke. Optical smoke alarms are therefore recommended in principal habitable rooms, however, if the room is used by a heavy smoker, this could give rise to some false alarms from tobacco smoke.

In cases where a principal habitable room is open plan with a kitchen, an optical smoke alarm is recommended to reduce the amount of unwanted alarms from cooking fumes.

Circulation spaces – most unwanted alarms occur during cooking. Optical smoke alarms are less sensitive from fumes caused by toasting bread or frying or grilling food. Therefore, optical smoke alarms are recommended in hallways and stairwells adjacent to kitchens.

2.11.5 Ionisation smoke alarms – domestic

Ionisation smoke alarms should conform to BS EN 14604:2005 and operate on the principle that the electrical current flowing between electrodes in an ionisation chamber is reduced when smoke particles enter the chamber. Ionisation smoke alarms are more sensitive to smoke containing small particles such as rapidly burning, flaming fires but are less sensitive to steam. Therefore, ionisation smoke alarms are recommended in hallways and stairwells adjacent to bathrooms or shower rooms to reduce the amount of unwanted false alarms.

Circulation spaces – multi-sensor alarms are recommended in hallways and stairwells adjacent to bathrooms or shower rooms to reduce the amount of unwanted false alarms.

2.11.6 Multi sensor alarms – domestic

A multi-sensor alarm provides the early warning of fire and can significantly reduce the amount of unwanted false alarms in certain circumstances. See BS 5839: Part 6:2004 for more detailed information.

2.11.7 Heat alarms – domestic

Heat alarms conforming to BS 5446: Part 2:2003 have fixed-temperature elements and operate on the principle of responding to the temperature of the fire gases in the immediate vicinity of the heat alarm. Heat alarms are used where ambient temperatures are likely to fluctuate rapidly over a short period such as in kitchens and are less likely to produce false alarms. Elsewhere, heat alarms should not be used instead of smoke alarms to reduce unwanted false alarms.

2.11.8 Siting of fire detectors – domestic

The guidance in this clause takes account of the audibility levels in adjoining rooms and the effect of smoke travelling along a ceiling.

Smoke alarms and heat alarms, by their definition, include an integral sounder. Smoke alarms are designed to produce a sound output of 85 dB(A) at 3 m.

Therefore, allowing for a sound attenuation through a domestic door, a sound level of between 55–65 dB(A) can be expected at the bed-head in each bedroom which should rouse the occupants. There is no evidence to suggest that lives are being lost in dwellings due to audibility levels other than when people are incapacitated to such a degree, for example by alcohol or drugs, that even higher sound levels would not waken them.

Smoke from a fire in a dwelling is normally hot enough that it rises and forms a layer below the ceiling. As the smoke rises and travels horizontally it mixes with air which increases the size of the smoke particles. This means that ionisation smoke alarms may be less sensitive to the smoke. Where a hallway is very long, the smoke might cool to such an extent that it loses buoyancy and spreads along the floor.

Audibility – therefore, smoke alarms should be located in circulation spaces:

(a) not more than 7 m from the door to a living room or kitchen
(b) not more than 3 m from every bedroom door and
(c) in circulation spaces more than 7.5 m long, no point within the circulation space should be more than 7.5 m from the nearest smoke alarm.

Smoke travel – a smoke alarm in the principal habitable room should be sited such that no point in the room is more than 7.5 m from the nearest smoke alarm and, in the case of a heat alarm, no point in the kitchen should be more than 5.3 m from the nearest heat detector.

All dimensions should be measured horizontally.

Smoke might not reach a smoke alarm where it is located on or close to a wall or other obstruction. Therefore, smoke alarms should be ceiling mounted and positioned away from any wall or light fitting. In order to reduce unwanted false alarms, smoke alarms should not be sited directly above heaters, air conditioning ventilators or other ventilators that might draw dust and fine particles into the smoke alarm.

Smoke alarms and heat alarms should be ceiling mounted and located such that their sensitive elements are:

(a) in the case of a smoke alarm, between 25 mm and 600 mm below the ceiling, and at least 300 mm away from any wall or light fittings and
(b) in the case of a heat alarm, between 25 mm and 150 mm below the ceiling.

2.11.9 Grade of fire detection and fire alarm system – domestic

The monitoring of wiring ensures that faults are alerted immediately and, hence, the time in which the system is disabled will be at a minimum.

Therefore, at least a Grade D fire detection and fire alarms system should be installed in every dwelling which comprises 1 or more mains powered smoke alarm and 1 or more mains powered heat alarm with an integral standby supply in accordance with BS 5839: Part 6:2004.

However a sheltered housing complex normally provides accommodation for vulnerable occupants with a diverse range of support needs. Therefore, a fire alarm signal should be transmitted to a remote monitoring service or to a warden who can assist with any evacuation if necessary, or call for assistance.

In order to achieve this principle, a Grade C system should be installed in every dwelling in a sheltered housing complex which comprises central control equipment in accordance with BS 5839: Part 6:2004 and

(a) 1 or more mains powered smoke alarms and 1 or more mains powered heat alarms with an integral standby supply or
(b) point fire detectors and separate sounders.

2.11.10 Wiring and power – domestic

Smoke alarms and heat alarms should be mains operated and permanently wired to a circuit which should take the form of either:

(a) an independent circuit at the main distribution board, in which case no other electrical equipment should be connected to this circuit (other than a dedicated monitoring device installed to indicate failure of the mains supply to the alarms) or

(b) a regularly used local lighting circuit that is separately electrically protected.

The standby supply for smoke alarms and heat alarms may take the form of a primary battery, a secondary battery or a capacitor.

The capacity of the standby supply should be sufficient to power the smoke alarms and heat alarms in the quiescent mode for at least 72 hours whilst giving an audible or visual warning of power supply failure, after which there should remain sufficient capacity to provide a warning for a further 4 minutes or, in the absence of a fire, a fault warning for at least 24 hours.

Interconnection – all smoke alarms and heat alarms in a dwelling should be interconnected so that detection of a fire in any alarm, operates the alarm signal in all of them. Smoke alarms and heat alarms should be interconnected in accordance with BS 5839: Part 6:2004.

The system should be installed in accordance with the manufacturer's written instructions. This should include a limitation on the number of smoke alarms and heat alarms which may be interconnected.

2.11.11 Radio linked systems – domestic

Radio-linked interconnection between hard wired smoke alarms and/or heat alarms may be used for a Grade D system. More detailed guidance on the use of radio-linked technology can be obtained from BS 5839: Part 6:2004.

2.11.0 Non-domestic

It is important that any outbreak of fire in premises is detected at an early stage in order that the occupants, once alerted, can commence evacuation of the premises as soon as possible. There should also be a means by which anyone in the building who discovers a fire can alert others to its existence. This should include arrangements for calling the fire and rescue service.

Risk assessment fire warning – in small single-storey non-residential buildings the means of raising the alarm could be quite simple, for example where a shouted warning "FIRE" by the person discovering the fire may be all that is needed. In more complex buildings, a sophisticated fire detection and fire alarm system may be needed.

False alarms – around 97 % of all automatic calls received by the fire and rescue service result in unnecessary attendance due to false alarms. This is normally attributed to poor design, installation or maintenance of automatic fire detection and alarm systems. Guidance on how to assess the risks and reduce false alarms is provided in BS 5839: Part 1:2002.

Additional guidance on special fire precautions that may be necessary within residential care buildings, hospitals and enclosed shopping centres is grouped in the annexes to the non-domestic Technical Handbook as follows:

(a) residential care buildings, see Annex 2.A
(b) hospitals, see Annex 2.B
(c) enclosed shopping centres, see Annex 2.C.

The guidance in the Annexes is in addition to the guidance to Standard 2.1 to 2.15.

2.11.1 Non-domestic to 2.11.14

Sections 2.11.1 to 2.11.14 (inclusive) of the Non-domestic Technical Handbook cover the following matters in some detail in relation to fire detection and fire alarm systems.

(a) 2.11.1 Evacuation methods used for the premises.
(b) 2.11.2 Assessment of the determined use of the premises.
(c) 2.11.3 Categories of fire detection and fire alarm system.
(d) 2.11.4 Residential care buildings.
(e) 2.11.5 Hospitals.
(f) 2.11.6 Shared residential accommodation.
(g) 2.11.7 Residential buildings (other than residential care buildings and hospitals).
(h) 2.11.8 Entertainment buildings and assembly buildings.
(i) 2.11.9 Offices and shops.
(j) 2.11.10 Educational building.
(k) 2.11.11 Factory buildings and storage buildings.
(l) 2.11.12 Enclosed shopping centres.
(m) 2.11.13 Transportation Terminals.
(n) 2.11.14 Other non-residential buildings.

4.5   Electrical safety

Domestic and non-domestic

Every building must be designed and constructed in such a way that the electrical installation does not:

(a) threaten the health and safety of the people in and around the building; and
(b) become a source of fire.

Limitation:

This standard does not apply to an electrical installation:

(a) serving a building or any part of a building to which the Mines and Quarries Act 1954 or the Factories Act 1961 applies; or
(b) forming part of the works of an undertaker to which regulations for the supply and distribution of electricity are made under the Electricity Act 1989.

4.5.0   Domestic and non-domestic

The hazards posed by unsafe electrical installation are injuries caused by contact with electricity (shocks and burns) and injuries arising from fires in buildings ignited through malfunctioning or incorrect installations.

The intention of this standard is to ensure that electrical installations are safe in terms of the hazards likely to arise from defective installations, namely fire, electric shock and burns or other personal injury.

Installations should:

(a) safely accommodate any likely maximum demand;

(b) incorporate appropriate automatic devices for protection against overcurrent or leakage; and

(c) provide means of isolating parts of the installation or equipment connected to it, as are necessary for safe working and maintenance.

The standard applies to fixed installations in buildings. An installation consists of the electrical wiring and associated components and fittings, including all permanently secured equipment, but excluding mobile (portable) equipment and appliances.

Index

A

Agricultural premises	Chap 2; 8.2
additional requirements for ELV circuits	8.2.4
luminaires in	8.2.6
radiant heaters	8.2.2
RCDs as a measure for protection against fire	8.2.3
risks of fire in	8.2.1
rodent activity in	8.2.5
Alarm systems	4.2; 7.2; 7.4; 7.6
Amusement devices	Chap 2; 8.7
Approved Document B	1.2.9; Appx B; Appx C; Appx D
Arcs	3.5
Arc welding, electric	3.5

B

British Standards (and BS ENs) mentioned in this Guidance Note	Appx A
Building (Inner London) Regulations 1985	1.2.14
Building Regulations 2000 (England and Wales)	1.2.9
Building Regulations (Northern Ireland) 2000 as amended	1.2.11
Building (Scotland) Regulations 2004 as amended	1.2.10
Burns, protection against	Chap 5

C

Cables	
heating	3.2.1
in locations with risk of fire due to nature of processed/ stored materials	4.3.4
in premises mainly made of combustible materials	4.4.3
plastic clips etc. not permitted	3.11; 4.2; 7.6
suitability for ambient temperature	Chap 2

Cinematograph (Safety) Regulations 1955 — 1.2.8

Circuit integrity — Chap 2; 4.2; 7.6

Circuits

    in locations with risk of fire due to nature of processed/
    stored materials — 4.3.4

Circuses — Chap 2

Classification of external influences — 4.3

Combustible constructional materials — 4.4

Concentration of heat — 3.7; Fig 3.4

Connections — 3.10; 8.6

Construction (Design and Management) Regulations 2015 — 1.2.5

Consumer units and similar switchgear assemblies — 3.9.1

Controlgear — 4.2; 4.3.3

D

Danger, definition of — 1.2

Dangerous Substances and Explosive Atmospheres
    Regulations 2002 — 1.2.6

Design — Introduction

Downlighters — 3.4
    Fig 3.2

Dust — 4.3; 4.3.2; 4.3.3;
    8.2.6

E

Electricity at Work Regulations 1989 — 1.2.1; 3.5

Electricity Safety, Quality and Continuity Regulations 2002 — 1.2.2; 4.3.4

Enclosures — 3.9; 4.3.2

    of consumer units and similar — 3.9.1

Equipment

    containing flammable liquid — 3.8

    installed in buildings of combustible construction — 4.4.1

Escape routes, requirements for — 4.2

Exhibitions, shows and stands — 8.6

Explosion risk — 4.3

External sealing arrangements — 6.2.1

Extra-low voltage circuits, additional requirements for in
    agricultural and horticultural premises — 8.2.4

Extra-low voltage lighting — 3.2.3; 8.5

F

Fairgrounds — Chap 2; 8.7

Fire detection and fire alarm systems — Appx B; Appx E

Fire propagating structures — 4.5

Fire risk

    due to nature of processed or stored materials — 4.3

    none in normal use — 3.4

Fire-segregated compartment, precautions in — 6.1

Fire-stopping — 4.5; 6.2.5; Appx D; Appx E

    external sealing arrangements — 6.2.1

    internal sealing arrangements — 6.2.2

Fixed equipment — 3.2; 3.3

Flammable liquid, equipment containing — 3.8; 4.2

Floor heating systems — Chap 2; 8.8

Floor warming cable — 3.2.1

Focusing of heat — 3.7; Fig 3.4

G

No entries

H

Hand-held part — Chap 5

Harmful effects — 3.1

Health and Safety at Work etc. Act 1974 — Introduction

Health and Safety (Safety Signs and Signals) Regulations 1996 — 1.2.12

Heater of liquid or other substance — Chap 5

Heating appliances — 4.3.6

Heating cables — 3.2.1

Heat storage appliances — 4.3.6

Horticultural premises (see also Agricultural premises) — Chap 2; 8.2

Hot particles — 3.5

HSR25 — 1.2.1

I

Ignition hazard, precautions to prevent — 4.4.1

Index of protection (IP) system — 4.3.3

Insulation, support of thermal — Fig 3.3

Internal fire spread — Appx C; Appx D; Appx E

Internal sealing arrangements — 6.2.2

J

Joints   3.10

K

No entries

L

Lampholders, T class   3.4
Licensed premises   Preface; 1.2.14
Lighting, extra-low voltage   3.2.3
Locations of national, commercial, industrial or public significance   4.6
London Building Acts (Amendment) Act 1939   1.2.14
Luminaires   3.2.2
    for use in agricultural or horticultural premises   8.2.6
    minimum distances from combustible constructional materials   4.4.2
    minimum distances from combustible processed or stored
    materials   4.3.1
    symbols for   3.2.4

M

Management of Health and Safety at Work Regulations 1999   1.2.3
Manufacturers' instructions   3.3
Medical locations   8.3
Mobile equipment   3.2; Appx E
Motors   4.3.5

N

No entries

O

Oil-filled equipment   3.8
Operational manual   Introduction
Overheating   Chap 2
    of luminaires installed in stands   8.6
    of terminations   3.10

P

Particular risks of fire, precautions   Chap 4
Penetrations, sealing of wiring system   6.2
Petroleum (Consolidation) Act 1928   1.2.7
Plastic cable clips etc. not permitted   4.2; 7.6
Portable equipment   3.2; Appx E
Precautions in a fire-segregated compartment   6.1

Product standards   Appx A

Protection against burns   Chap 5

Provision and Use of Work Equipment Regulations
1998   1.2.4

Q

No entries

R

Radiant heaters   3.7; 8.2.2

RCDs, as measure of protection against fire   Chap 2; 4.3.4

Regulatory Reform (Fire Safety) Order 2005   1.2.13

Rodent activity (agricultural and horticultural premises)   8.2.5

S

Safety services   Chap 7

    changeover times   7.2

    circuits for   7.4

    classification   7.2

    general   7.1

    generators used as source for   7.3.1

    information to be made available   7.5

    sources for   7.3

    standards   Appx A

    wiring systems   7.6

Scottish Technical Standards   Appx E

Sealing of wiring system penetrations   6.2

    external sealing arrangements   6.2.1

    internal sealing arrangements   6.2.2

Section 20 London Building Acts (Amendment) Act
1939   1.2.14

Solar photovoltaic (PV) power systems   8.4

Sparks   3.5

Special installations and locations, protection against
fire in   Chap 8

Specification   Introduction

Spotlights   3.7

Standards of relevance   Appx A

Statutory Regulations   Chap 1

Surface heating systems   3.2.1

Switchgear   4.2; 4.3.3

Symbols used in connection with luminaires,
controlgear etc.   3.2.4

T

Temperature, maximum, of accessible surfaces   Table 5.1

Terminations   3.10

Thermal imaging   3.10

Thermal insulation, support box for   Fig 3.3

Thermal cut-out devices   Chap 2; 4.1

U, V

No entries

W

Welding sets, electric arc   3.5

Wiring systems

    in escape routes   3.11

    in locations with risk of fire due to nature of processed/stored
    materials   4.3.4

    sealing of penetrations   6.2

X, Y, Z

No entries

Guidance Note 4: Protection Against Fire
© The Institution of Engineering and Technology

ENDORSED BY

City & Guilds

Online training for BS 7671:2018

With the IET's interactive online training from the IET Academy, there's no need to take time away from the job to train for your C&G 2382:18 (or equivalent) qualification.

Book a full or Update course and get:

- **Courses that have been endorsed by C&G, ensuring the best quality content**
- **Learning at your own pace with 12 months access to all content**
- **Online/offline access to suit you, through our app**
- **Three full practice exams to fully prepare you for the real thing**
- **A certificate of completion to present to your local exam centre to sit your C&G 2382:18 exam**

Find out more and book your course now at: www.theiet.org/ academy-regsa